Engineering Maths for Applications

J.S. Han

O-Sung Media

Preface

College students majoring in electronics engineering are expected to acquire a great amount of mathematical knowledge via engineering courses, such as engineering mathematics, random variable and stochastic process, etc, as well as primary courses, such as general mathematics and linear algebra. However, these approaches are usually oriented in mathematics that even though all the theories are understood, there always have been lack of capability among students to apply these theories to the practical engineering fields. This book covers the theories for the fundamental electronics engineering fields, such as circuit analysis, electromagnetic, communication theory and digital signal processing. This book is different from the other texts of engineering mathematics by reconstructing theories analyzed in the aspects of engineering fields. Breaking from the conventional approaches in which students should study mathematical theories, and then, apply them to the engineering fields of study, this book attempts to save time and effort of students by having them study the major fields and then acquire necessary mathematical theories concerning. This book consists of two parts; part 1 covers vector analysis, circuit analysis, signal analysis (including communication) and electromagnetic field theory; part 2 covers differential equation, complex number, Laplace transform and Fourier series. The electromagnetic engineering fields in Part 1 and the mathematical theories in Part 2 are not original works of the author, but are general knowledge widely used in related areas. In my experience, lectures based on the approach adopted in this book are highly effective in enhancing the mathematical capabilities of students. This book is published thanks to Mr. S.G. Oh, CEO of O-Sung Media. I expect that this book will be useful for many students who are not interested in electronics engineering because of the mathematics. In addition, I am thankful to O-Sung Media for their efforts in collecting necessary data. The author appreciates many people's encouragement in writing this book, especially advice of K.P.U staffs.

I wish to thank my son and wife.

J.S. Han

CONTENTS

Part I

Part II

Chapter 5

Chapter 6

Part I

Chapter 1

Vector Analysis of Signal

In representing the physical quantity of an object, scalar quantity requires only its magnitude (or length), while a vector quantity requires both the magnitude and direction. Length, speed, mass, angle and temperature are scalar quantities, while displacement, velocity, acceleration and forces are vector quantities.

Vector $\mathbf{u_A}$ maintaining the direction of a vector A and having a length of one is called a unit vector.

$$\mathbf{A} = A\mathbf{u}_A$$

where, A indicates the magnitude of vector A, and is represented as $A = |\mathbf{A}|$.

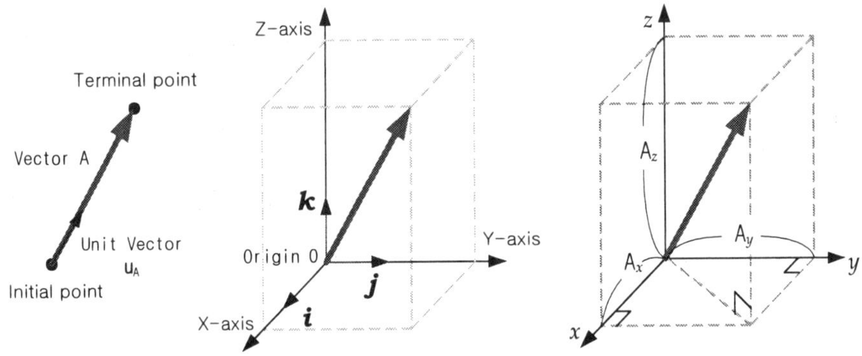

Figure 1-1 Representations of vector

That is $u_A = \frac{A}{|A|} = \frac{A}{A}$ and $|u_A| = 1$.

The vector with the length of 1 in the direction of a reference axis of a coordinate system is called a basis vector. On a Cartesian coordinate system, the basis vectors in the incremental direction of x-axis, y-axis, and z-axis are $\boldsymbol{i}, \boldsymbol{j}$, and $\boldsymbol{k}$, respectively.

Where vector projections of axes x, y and z for vector A are A_x, A_y and A_z, respectively, the vector can be represented as the following with the basis vectors.

$$A = \boldsymbol{i}A_x + \boldsymbol{j}A_y + \boldsymbol{k}A_z$$
$$= A_x\boldsymbol{i} + A_y\boldsymbol{j} + A_z\boldsymbol{k}$$
$$|A| = \sqrt{A_x^2 + A_y^2 + A_z^2}$$

In Figure 1-2, a parallelogram is drawn with vectors A and B having the same origin. OP, one of the diagonals of the parallelogram, is calculated with A+B.

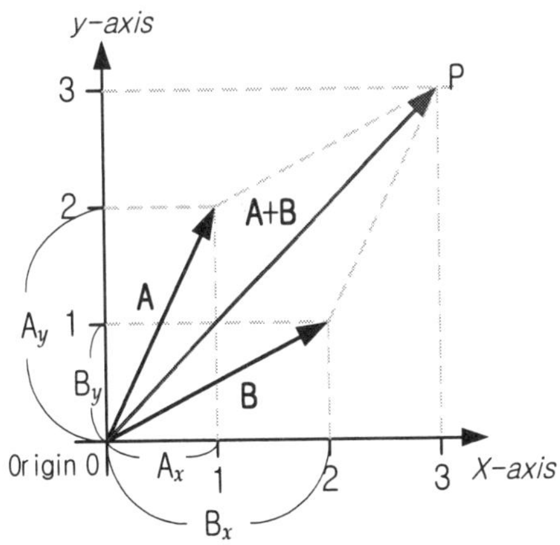

Figure 1-2 Parallelogram rule

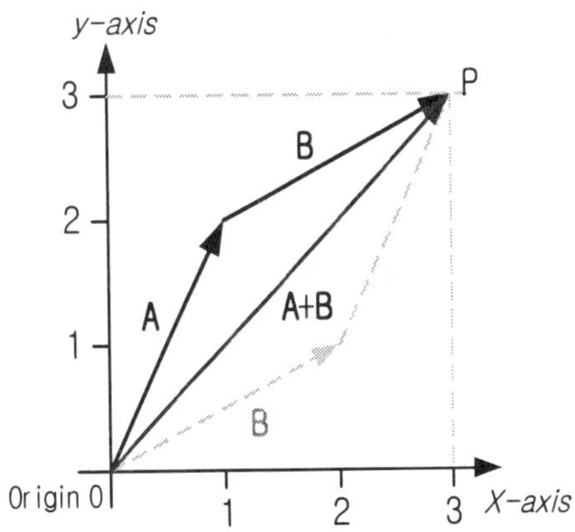

Figure 1-3 Triangle method

In Figure 1-3, the adding Vector B is transferred in parallel so that the start of Vector B should be placed on the endpoint of the Vector A. The base line OP of the triangle is calculated with the sum of vectors A+B.

$$|S| = |A+B| = |\boldsymbol{i}(A_x+B_x)+\boldsymbol{j}(A_y+B_y)+\boldsymbol{k}(A_z+B_z)|$$
$$= \sqrt{(A_x+B_x)^2+(A_y+B_y)^2+(A_z+B_z)^2}$$

$$A-B=(\boldsymbol{i}A_x+\boldsymbol{j}A_y+\boldsymbol{k}A_z)-(\boldsymbol{i}B_x+\boldsymbol{j}B_y+\boldsymbol{k}B_z)$$
$$=\boldsymbol{i}(A_x-B_x)+\boldsymbol{j}(A_y-B_y)+\boldsymbol{k}(A_z-B_z)$$
$$=-(B-A)$$

In Figure 1-4, an additional parallelogram is drawn with the negative(-) vector having the same length with a vector but in the opposite direction (180°).

Figure 1-5 shows the subtraction of two vectors. To subtract vector B from vector A, place the origins of A and B at the same point, and then draw an arrow from the endpoint of B to the endpoint of A. That arrow represents the vector A-B.

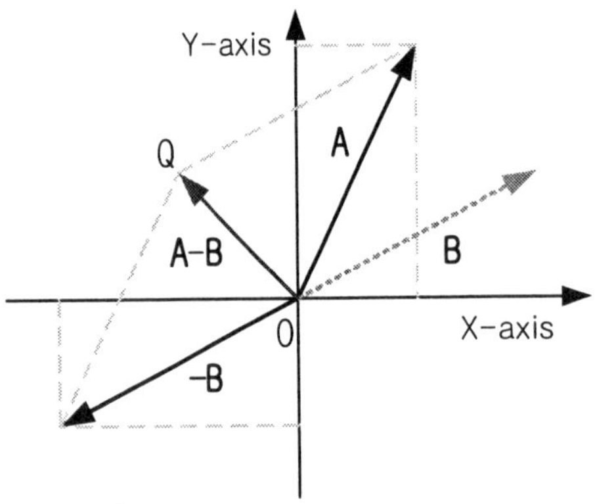

Figure 1-4 Parallelogram rule

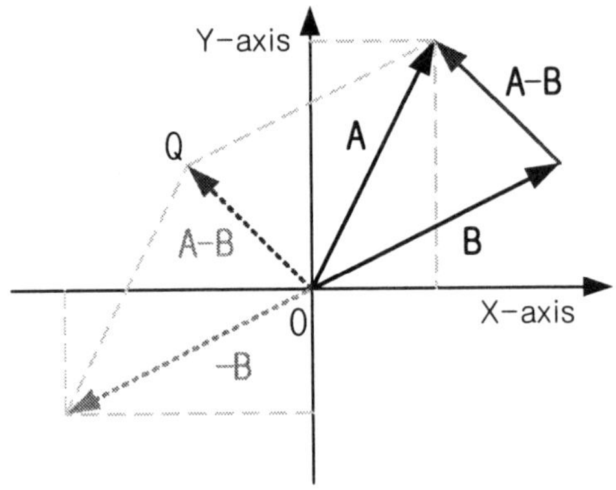

Figure 1-5 Triangle method

The magnitude of the vector is calculated as follows:

$$|D| = |A-B| = |\boldsymbol{i}(A_x - B_x) + \boldsymbol{j}(A_y - B_y) + \boldsymbol{k}(A_z - B_z)|$$

$$= \sqrt{(A_x - B_x)^2 + (A_y - B_y)^2 + (A_z - B_z)^2}$$

The magnitude of a vector is changed if the vector is multiplied by a scalar. The following formula shows vector A multiplied by scalar a:

$$A = \boldsymbol{i}A_x + \boldsymbol{j}A_y + \boldsymbol{k}A_z, a: \text{ Real number}$$

$$A \times a = \mathbf{A} \cdot a = Aa = aA$$

$$a\mathbf{A} = \boldsymbol{i}(aA_x) + \boldsymbol{j}(aA_y) + \boldsymbol{k}(aA_z)$$

$$= \boldsymbol{i}(A_x a) + \boldsymbol{j}(A_y a) + \boldsymbol{k}(A_z a)$$

$$|a\mathbf{A}| = \sqrt{(aA_x)^2 + (aA_y)^2 + (aA_z)^2}$$

The inner product of vector A and vector B is indicated with a dot, and the result of the calculation is a scalar. Where the angle between the vectors A and B with the same origin is θ, the following formula is made:

$$A \cdot B = |A| \cdot |B| \cos\theta = |\boldsymbol{B}| \cdot |\boldsymbol{A}| \cos\theta = \boldsymbol{B} \cdot \boldsymbol{A}$$

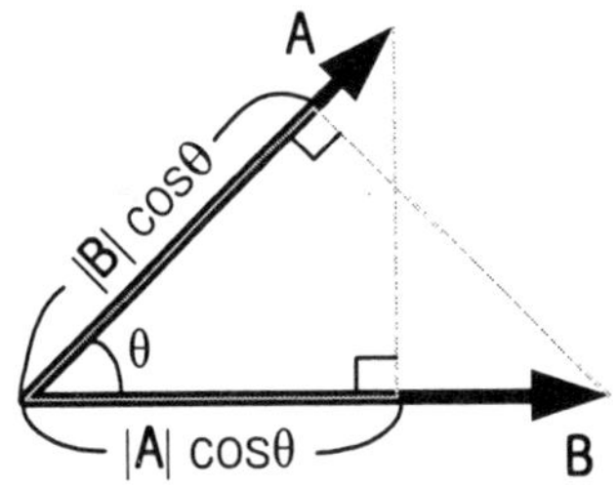

Figure 1-6 Inner product of vector

On a Cartesian coordinate system, the inner product between basis vectors is defined as follows:

$$\boldsymbol{i}\cdot\boldsymbol{i}=\boldsymbol{j}\cdot\boldsymbol{j}=\boldsymbol{k}\cdot\boldsymbol{k}=1\cdot 1\cdot\cos 0^{0}=1$$

$$\boldsymbol{i}\cdot\boldsymbol{j}(\boldsymbol{j}\cdot\boldsymbol{i})=\boldsymbol{j}\cdot\boldsymbol{k}(\boldsymbol{k}\cdot\boldsymbol{j})=\boldsymbol{k}\cdot\boldsymbol{i}(\boldsymbol{i}\cdot\boldsymbol{k})=1\cdot 1\cdot\cos 90^{o}=0$$

Using these characteristics, the inner product of vectors is calculated on the Cartesian coordinate system as follows:

$$\begin{aligned}
\mathrm{A}\cdot\mathrm{B}&=(\boldsymbol{i}\mathrm{A}_x+\boldsymbol{j}\mathrm{A}_y+\boldsymbol{k}\mathrm{A}_z)\cdot(\boldsymbol{i}\mathrm{B}_x+\boldsymbol{j}\mathrm{B}_y+\boldsymbol{k}\mathrm{B}_z)\\
&=\boldsymbol{i}\cdot\boldsymbol{i}(\mathrm{A}_x\mathrm{B}_x)+\boldsymbol{j}\cdot\boldsymbol{j}(\mathrm{A}_y\mathrm{B}_y)+\boldsymbol{k}\cdot\boldsymbol{k}(\mathrm{A}_z\mathrm{B}_z)\\
&\quad+\boldsymbol{i}\cdot\boldsymbol{j}(\mathrm{A}_x\mathrm{B}_y)+\boldsymbol{i}\cdot\boldsymbol{k}(\mathrm{A}_x\mathrm{B}_z)+\boldsymbol{j}\cdot\boldsymbol{i}(\mathrm{A}_y\mathrm{B}_x)\\
&\quad+\boldsymbol{j}\cdot\boldsymbol{k}(\mathrm{A}_y\mathrm{B}_z)+\boldsymbol{k}\cdot\boldsymbol{i}(\mathrm{A}_z\mathrm{B}_x)+\boldsymbol{k}\cdot\boldsymbol{j}(\mathrm{A}_z\mathrm{B}_y)\\
&=(\mathrm{A}_x\mathrm{B}_x)+(\mathrm{A}_y\mathrm{B}_y)+(\mathrm{A}_z\mathrm{B}_z)=\mathrm{B}\cdot\mathrm{A}
\end{aligned}$$

Outer product of vectors A and B is indicated as a cross product, and the result of the calculation is another vector. Where the angle between the vectors with the same origin is θ, the following formula is made:

$$A \times B = P \begin{cases} \text{Length : } |A| \cdot |B| \sin\theta \\ \text{The area of the parallelogram having 2 vectors as adjacent sides} \\ \text{Direction : Right handed direction in the order of multiplication after} \\ \text{gathering the origins to a point.} \end{cases}$$

$= -B \times A$

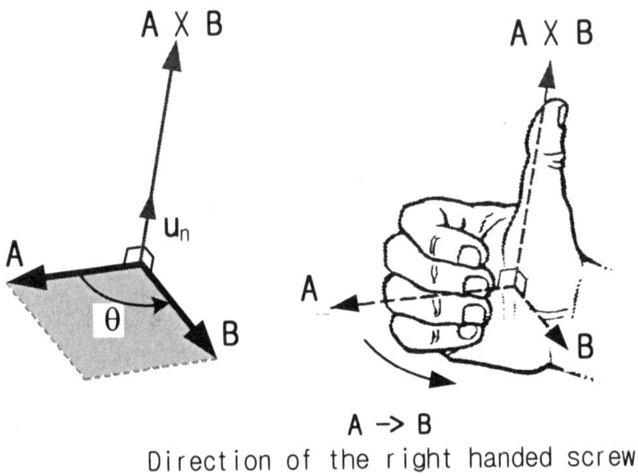

Figure 1-7 Outer product of vector

On a Cartesian coordinate system, the outer product between basis vectors is defined as follows:

$$\boldsymbol{i} \times \boldsymbol{i} = \boldsymbol{j} \times \boldsymbol{j} = \boldsymbol{k} \times \boldsymbol{k} = 0$$
$$\boldsymbol{i} \times \boldsymbol{j} = \boldsymbol{k}, \boldsymbol{j} \times \boldsymbol{i} = -\boldsymbol{k}$$
$$\boldsymbol{j} \times \boldsymbol{k} = \boldsymbol{i}, \boldsymbol{k} \times \boldsymbol{j} = -\boldsymbol{i}$$
$$\boldsymbol{k} \times \boldsymbol{i} = \boldsymbol{j}, \boldsymbol{i} \times \boldsymbol{k} = -\boldsymbol{j}$$

Using these characteristics, the outer product of vectors is calculated on the Cartesian coordinate system as follows:

$$\mathrm{A}\times\mathrm{B}=(\boldsymbol{i}\mathrm{A}_x+\boldsymbol{j}\mathrm{A}_y+\boldsymbol{k}\mathrm{A}_z)\times(\boldsymbol{i}\mathrm{B}_x+\boldsymbol{j}\mathrm{B}_y+\boldsymbol{k}\mathrm{B}_z)$$

$$=\boldsymbol{i}(\mathrm{A}_y\mathrm{B}_z-\mathrm{A}_z\mathrm{B}_y)+\boldsymbol{j}(\mathrm{A}_z\mathrm{B}_x-\mathrm{A}_x\mathrm{B}_z)+\boldsymbol{k}(\mathrm{A}_x\mathrm{B}_y-\mathrm{A}_y\mathrm{B}_x)$$

$$=\begin{vmatrix}\boldsymbol{i} & \mathrm{j} & \mathrm{k}\\ \mathrm{A}_x & \mathrm{A}_y & \mathrm{A}_z\\ \mathrm{B}_x & \mathrm{B}_y & \mathrm{B}_z\end{vmatrix}=-\mathrm{B}\times\mathrm{A}$$

If a certain quantity varies depending on the location of a space, such space is referred to as a field for such quantity. Depending on whether the quantity is scalar or vector, it is a scalar field or a vector field. Therefore, it is required to define the spatial rate of change of a scalar field or vector field in a given time, and a vector differential operator is used for this purpose.

Vector differential operator is defined as the following with the partial differentiation and the basis vector of a Cartesian coordinate system, and is called 'del' or 'nabla':

$$\nabla=\boldsymbol{i}\frac{\partial}{\partial x}+\boldsymbol{j}\frac{\partial}{\partial y}+\boldsymbol{k}\frac{\partial}{\partial z}$$

Gradient calculates the magnitude and direction of the maximum increase rate in a space for a scalar field. The result of the calculation is a vector.

In a Cartesian coordinate system, the gradient operation for a scalar function $V(x,y,z)$ is as follows:

$$\nabla \mathrm{V}=\boldsymbol{i}\frac{\partial \mathrm{V}}{\partial x}+\boldsymbol{j}\frac{\partial \mathrm{V}}{\partial y}+\boldsymbol{k}\frac{\partial \mathrm{V}}{\partial z}$$

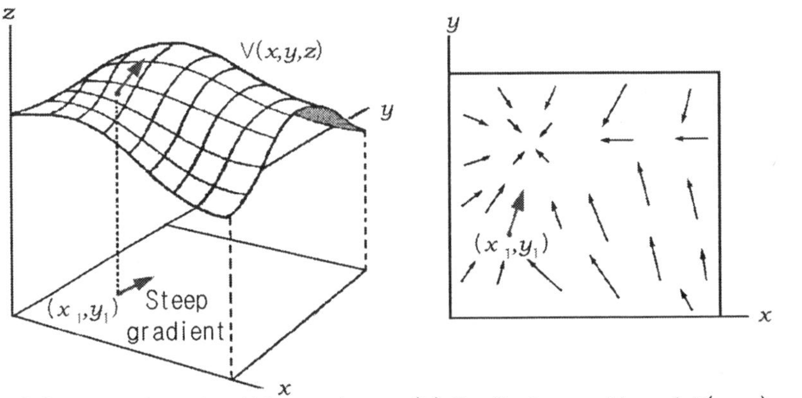

(a) Scalar function V(x, y, z) (b) Gradient operation of V(x, y)∇V

Figure 1-8 Gradient operation

Divergence measures the quantity of inflow or outflow when the differential volume gets close to 0 at a given point of a space for a vector field. The result of the calculation is a scalar. This result is indicated as the inner product between the vector differential operator and a vector function A.

In a Cartesian coordinate system, the divergence operation for the vector function $A = \boldsymbol{i}A_x + \boldsymbol{j}A_y + \boldsymbol{k}A_z$ is as follows:

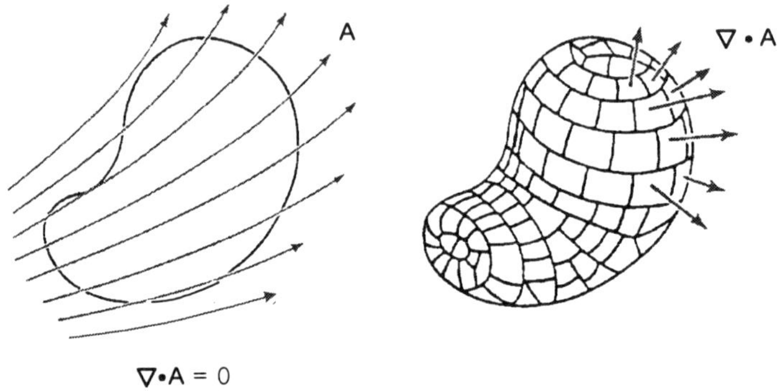

Figure 1-9 Divergence operation

$$\nabla \cdot \mathrm{A} = \left(\boldsymbol{i}\frac{\partial}{\partial x} + \boldsymbol{j}\frac{\partial}{\partial y} + \boldsymbol{k}\frac{\partial}{\partial z} \right) \cdot (\boldsymbol{i}\mathrm{A}_x + \boldsymbol{j}\mathrm{A}_y + \boldsymbol{k}\mathrm{A}_z)$$

$$= \frac{\partial \mathrm{A}_x}{\partial x} + \frac{\partial \mathrm{A}_y}{\partial y} + \frac{\partial \mathrm{A}_z}{\partial z}$$

For curl or rotation, a vortex in a vector field is created by the vector field that makes a surrounding rotation. The maximum quantity of circulation in the vector field around the closed path of the border of a differential area when the differential area of a vortex becomes close to 0, and the direction of circulation which is perpendicular to the area are calculated. The result of the calculation is a vector.

This result is represented as the outer product between the vector differential operator and a vector function A. In a Cartesian coordinate system, the rotation operation for the vector function $\mathrm{A} = \boldsymbol{i}\mathrm{A}_x + \boldsymbol{j}\mathrm{A}_y + \boldsymbol{k}\mathrm{A}_z$ is as follows:

$$\nabla \times \mathrm{A} = \left(\boldsymbol{i}\frac{\partial}{\partial x} + \boldsymbol{j}\frac{\partial}{\partial y} + \boldsymbol{k}\frac{\partial}{\partial z} \right) \times (\boldsymbol{i}\mathrm{A}_x + \boldsymbol{j}\mathrm{A}_y + \boldsymbol{k}\mathrm{A}_z)$$

$$= \begin{vmatrix} \boldsymbol{i} & \boldsymbol{j} & \boldsymbol{k} \\ \frac{\partial}{\partial x} & \frac{\partial}{\partial y} & \frac{\partial}{\partial z} \\ \mathrm{A}_x & \mathrm{A}_y & \mathrm{A}_z \end{vmatrix}$$

$$= \boldsymbol{i}\left(\frac{\partial \mathrm{A}_z}{\partial y} - \frac{\partial \mathrm{A}_y}{\partial z} \right) + \boldsymbol{j}\left(\frac{\partial \mathrm{A}_x}{\partial z} - \frac{\partial \mathrm{A}_z}{\partial x} \right) + \boldsymbol{k}\left(\frac{\partial \mathrm{A}_y}{\partial x} - \frac{\partial \mathrm{A}_x}{\partial y} \right)$$

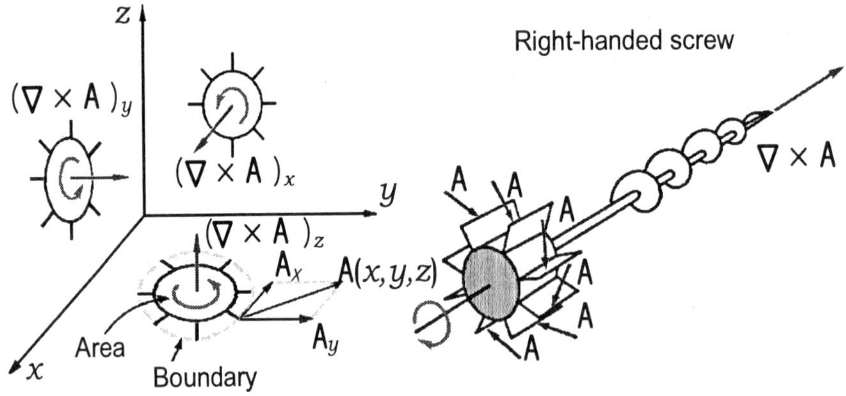

Figure 1-10 Rotation operation

For a scalar function $V(x,y,z)$, a vector double differentiation in the Cartesian coordinate system is as follows. The vector double differentiation is denoted by DEL^2, which is called laplacian.

$$\nabla \cdot \nabla V = \left(\boldsymbol{i}\frac{\partial}{\partial x} + \boldsymbol{j}\frac{\partial}{\partial y} + \boldsymbol{k}\frac{\partial}{\partial z} \right) \cdot \left(\boldsymbol{i}\frac{\partial V}{\partial x} + \boldsymbol{j}\frac{\partial V}{\partial y} + \boldsymbol{k}\frac{\partial V}{\partial z} \right)$$

$$= \frac{\partial^2 V}{\partial x^2} + \frac{\partial^2 V}{\partial y^2} + \frac{\partial^2 V}{\partial z^2}$$

$$= \left(\boldsymbol{i}\frac{\partial}{\partial x} + \boldsymbol{j}\frac{\partial}{\partial y} + \boldsymbol{k}\frac{\partial}{\partial z} \right) \cdot \left(\boldsymbol{i}\frac{\partial}{\partial x} + \boldsymbol{j}\frac{\partial}{\partial y} + \boldsymbol{k}\frac{\partial}{\partial z} \right) V$$

$$= \left(\frac{\partial^2}{\partial x^2} + \frac{\partial^2}{\partial y^2} + \frac{\partial^2}{\partial z^2} \right) V = \nabla^2 V$$

A coordinate system is used to represent the given vector on a 3-dimensional space. Out of many types of coordinate systems, the Cartesian coordinate system is generally used. A Cartesian coordinate system consists of a vertical axis (x-axis), a horizontal axis (y-axis) and a height axis (z-axis), which satisfy mutually perpendicular.

Where the angle between vector A and the x-axis, y-axis and z-axis on a

Cartesian coordinate system are α, β, and γ, respectively, $\cos\alpha$, $\cos\beta$ and $\cos\gamma$ are direction cosines of vector A. A is the magnitude of vector A, and A_x, A_y, A_z are the component values projected perpendicularly to each axis.

A coordinate system is used to decide a point on a 3-dimensional space, or to decide the reference position and direction. Coordinate systems are classified into cartesian coordinate, cylindrical coordinate and spherical coordinate.

Differential coordinate systems decide a point with the intersection point between 3 sides. Therefore, this coordinate system is called the orthogonal coordinate system.

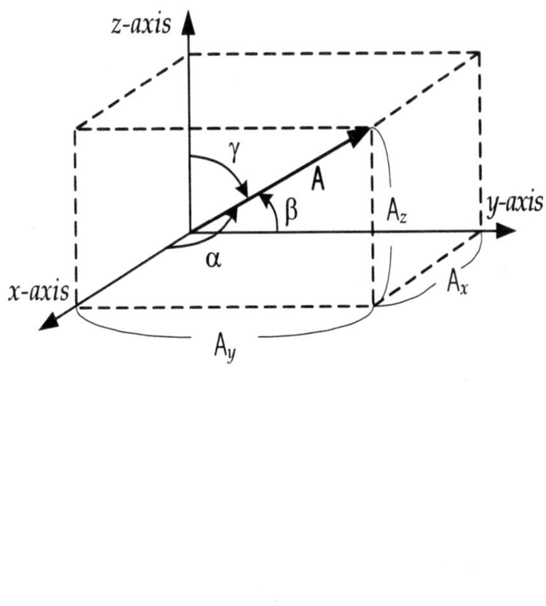

$$\cos\alpha = \frac{A_x}{A}$$

$$\cos\beta = \frac{A_y}{A}$$

$$\cos\gamma = \frac{A_z}{A}$$

1) Types of coordinate systems and notations

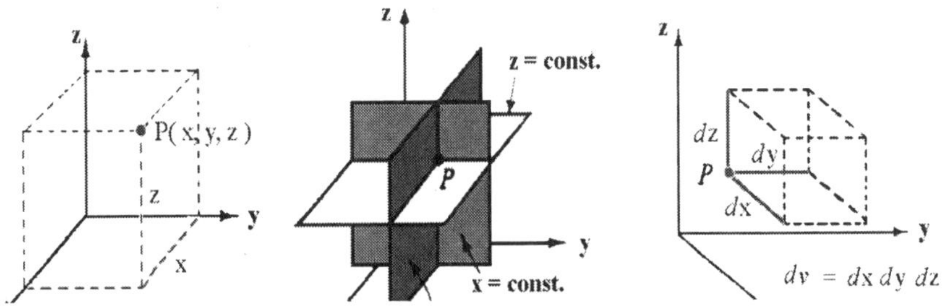

Figure 1-11 Cartesian coordinate system

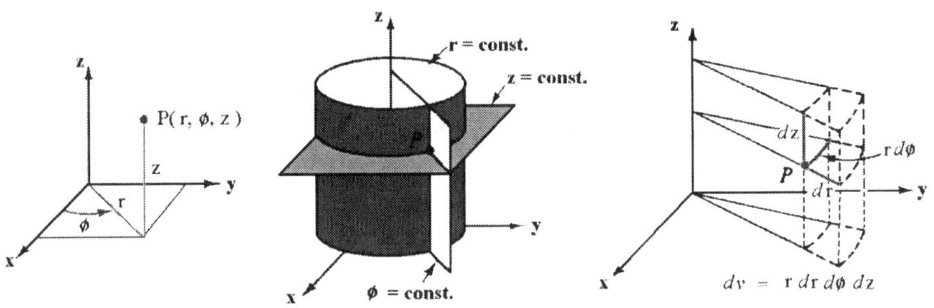

Figure 1-12 Cylindrical coordinate system

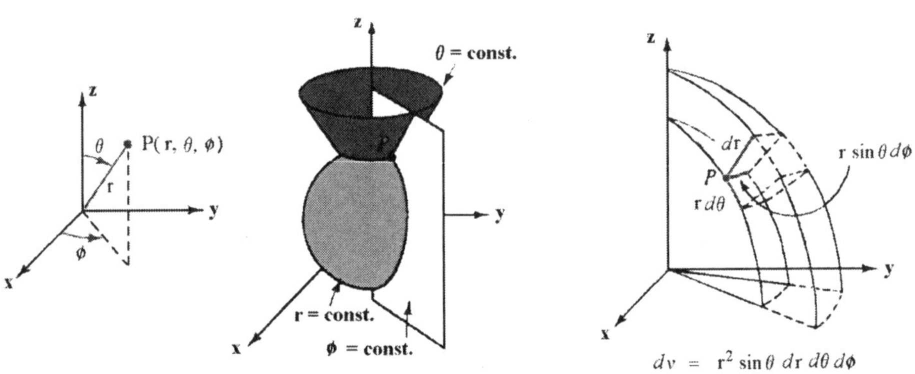

Figure 1-13 Spherical coordinate system

	Cartesian coordinate system	Cylindrical coordinate system	Spherical coordinate system
Coordinate axes	x, y, z	r, ϕ, z	r, θ, ϕ
Differential length	dx	dr	dr
	dy	$r \cdot d\phi$	$r \cdot d\theta$
	dz	dz	$r \cdot \sin\theta \cdot d\phi$
Differential area	$dx \cdot dy$	$r \cdot dr \cdot d\phi$	$r \cdot dr \cdot d\theta$
	$dy \cdot dz$	$r \cdot d\phi \cdot dz$	$r^2 \sin\theta \cdot d\theta \cdot d\phi$
	$dz \cdot dx$	$dz \cdot dr$	$r \cdot \sin\theta \cdot dr \cdot d\phi$
Differential volume	$dx \cdot dy \cdot dz$	$r \cdot dr \cdot d\phi \cdot dz$	$r^2 \cdot \sin\theta \cdot dr \cdot d\theta \cdot d\phi$

2) Conversion of coordinates

① Conversion between cartesian coordinate system and cylindrical coordinate system

Cylindrical coordinate system → Cartesian coordinate system	Cartesian coordinate system → Cylindrical coordinate system
$x = r \cdot \cos\phi$	$r = \sqrt{x^2 + y^2}$
$y = r \cdot \sin\phi$	$\phi = \tan^{-1}\left(\frac{y}{x}\right)$
$z = z$	$z = z$

② Conversion between cartesian coordinate system and spherical coordinate system

Spherical coordinate system → Cartesian coordinate system	Cartesian coordinate system → Spherical coordinate system
$x = r \cdot \sin\theta \cdot \cos\phi$	$r = \sqrt{x^2 + y^2 + z^2}$
$y = r \cdot \sin\theta \cdot \sin\phi$	$\theta = \cos^{-1}\left(\frac{z}{\sqrt{x^2 + y^2 + z^2}}\right)$
$z = r \cdot \cos\theta$	$\phi = \tan^{-1}\left(\frac{y}{x}\right)$

Exercise

Estimate cylindrical and spherical coordinates for the point $(1,\ 1, \sqrt{2})$:

Solution

① Cylindrical coordinate $r = \sqrt{x^2+y^2} = \sqrt{1^2+1^2} = \sqrt{2}$

② Cylindrical coordinate $\phi = \tan^{-1}\left(\frac{y}{x}\right) = \tan^{-1}(1) = 45^o$

③ Cylindrical coordinate $z =$ Cartesian coordinate system $z = \sqrt{2}$

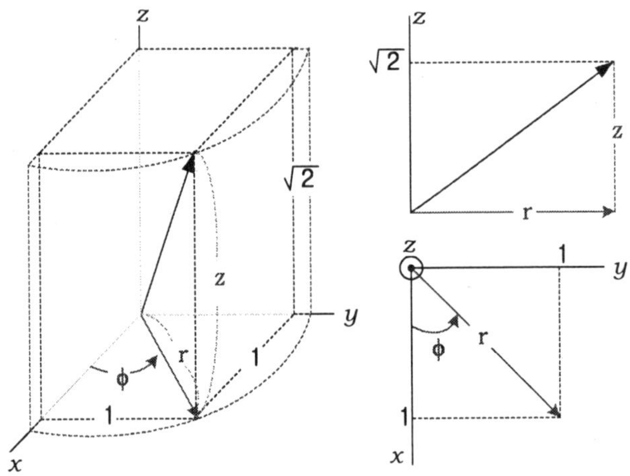

④ Spherical coordinate $r = \sqrt{x^2+y^2+z^2} = \sqrt{1^2+1^2+(\sqrt{2})^2} = 2$

⑤ Spherical coordinate

$$\theta = \tan^{-1}\left(\frac{\sqrt{x^2+y^2}}{z}\right) = \tan^{-1}\left(\frac{\sqrt{1^2+1^2}}{\sqrt{2}}\right) = \tan^{-1}(1) = 45^o$$

⑥ Spherical coordinates ϕ=Cylindrical coordinate

$$\phi = \tan^{-1}\left(\frac{y}{x}\right) = \tan^{-1}(1) = 45^o$$

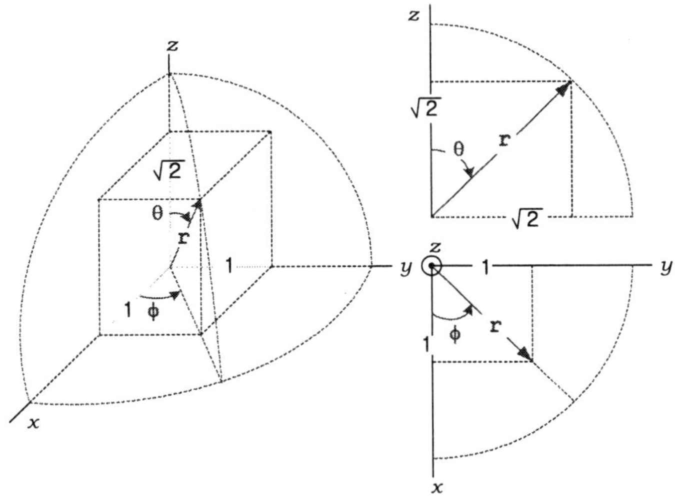
z
√2
θ
r
1
1 ϕ
y
x
z
√2
θ
r
√2
z
1
y
y
ϕ
r
1
x

Chapter 2

Circuit Analysis

The phenomenon that particles of a material electrified by an external electric field move in a space, or that free electrons move freely through conductive atoms in a conductor is called "electric current".

The direction of electric current is determined by the direction of the external electric field. Typically, the electric current generated in a conductor flows in the direction of the electric field. In reality, the flow of free electrons determines the electric current, and therefore, the direction of electric current is the complete opposite of the free electrons. Intensity of electric current is represented with the electric charges changed or moved in a given time.

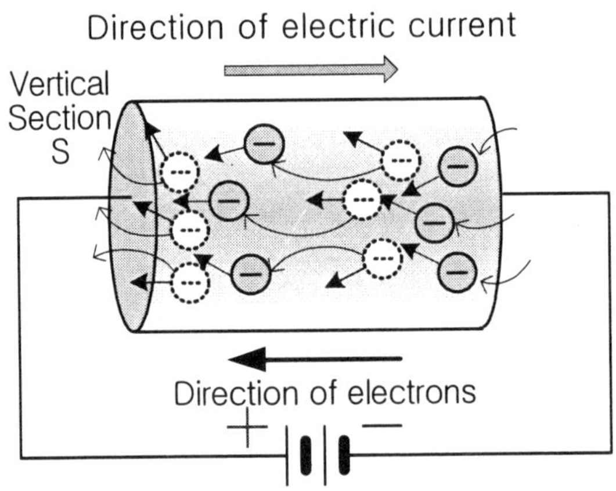

Figure 2-1 Electric current in a conductor

$$I = d\frac{Q(t)}{dt}[A] \quad \Rightarrow Q(t) = \int_{Q_1}^{Q_2} dQ = \int_{t_1}^{t_2} I \cdot dt[C]$$

Where Q is the electric charge change, and is represented with a time function. If electric charge change is constant over time, a constant electric current is maintained in a constant direction. This state is called DC (Direct Current). The state in which the quantity of electric charge increases or decreases over time and the direction of electric current changes is called AC (Alternating Current).

In the MKS unit system, the unit of electric current is Ampere [A]. If the quantity of electric charge moves for a second is 1 [C], the amount of electric current flown is 1 [A].

Current density is a vector amount, and is represented as the quantity an electric charge changes per a unit time (1[sec]) in a perpendicular unit area (1[m^2]).

Select a cross section which the current passes through perpendicularly, and draw a normal which is perpendicular to the cross section, and we will get an area vector with the area along the direction of the normal in the direction of electric current as the magnitude. The direction of the area vector is the direction of the current density vector. In the above figure, select a closed surface S inside a conductor in which electric current flows, and the electric current and the current density on S are expressed as the following:

$$J = n\frac{dI}{dS}[A/m^2] \quad \Rightarrow \quad I = \int_S J \cdot dS = \int_S J \cdot n dS[A]$$

Where, vector n is the unit vector with the magnitude of 1. If the electric current 1[A] moves by the area 1[m^2] perpendicularly, the magnitude of the current density vector is 1[A/m^2].

In a closed surface S, if we integrate the current density J in this closed surface, we will get the quantity of electric charge passing through the closed curve for a unit time, which is the electric current. Using Gauss' theorem, in the relational expression of electric current and current density, the integrated area can be replaced by the volume integration.

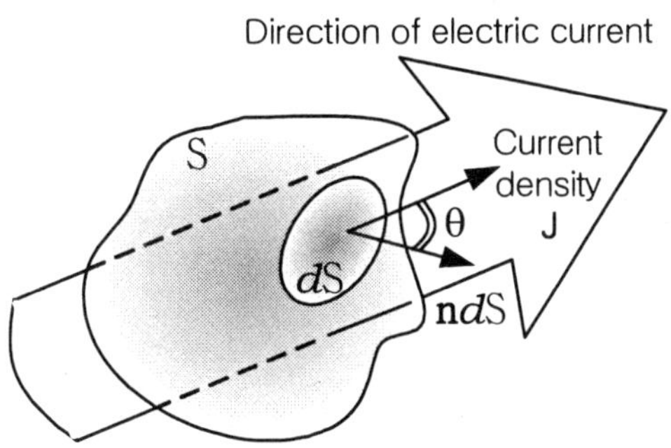

Figure 2-2 Current density

$$I=\oint_S \mathbf{J}\cdot \mathbf{n}dS=\int_v \nabla\cdot \mathbf{J}dv[A]$$

If the electric current I flows out from the closed surface, the quantity of electric charge in the differential volume is decreased. Where the charge density in the differential volume v is $\rho_v[C/m^3]$, the following formulas are made:

$$I=\oint_S \mathbf{J}\cdot \mathbf{n}dS=-\frac{\partial Q}{\partial t}=-\int_v \frac{\partial \rho_v}{\partial t}dv[A]$$

In the above two formulas, divergence of current density is calculated with time evolution of the volume charge.

$$\int_v \nabla\cdot \mathbf{J}dv=-\int_v \frac{\partial \rho_v}{\partial t}dv$$

$$\therefore \nabla\cdot \mathbf{J}=-\frac{\partial \rho_v}{\partial t}$$

As the quantity of electric charge is reduced with the time t, the electric current diverges. In other words, the quantity of electric charge flowing out through the closed surface S is equal to the quantity of electric charge reduced from the closed surface. For the normal electric current in which the charge density does not change at a point, divergence of the current density is 0. This is called a continuity of electric current.

If there is no change of electric current or quantity of electric charge, the divergence value is 0. Kirchhoff's Current Law , that "the sum of currents entering the node is equal to the sum of currents leaving the node", implies the continuity of current.

Electric resistance R varies depending on the shape and material of the conductor. Where the length of a conductor is $\ell[m]$ and the cross sectional area is $S[m^2]$, the following formula is made.

$$R = \frac{\rho \ell}{S} [\Omega]$$

Where, ρ is specific resistance or resistivity.

The reciprocal of resistance $\frac{1}{R}$ is conductance, and is represented as the following:

$$G = \frac{1}{R} = \frac{1}{\rho} \cdot \frac{S}{\ell} = \sigma \cdot \frac{S}{\ell} [\mho]$$

Where, ρ : Resistivity (Ωm)

σ : Conductivity ($\mho$/m)

ℓ : Length of conducting wire (m)

S : Cross sectional area of conducting wire (m^2)

1. Kirchhoff's first law

The algebraic sum of the currents entering any node is zero.

$$\sum_{i=1}^{n} I_i = 0$$

2. Kirchhoff's second law

The directed sum of the electrical potential differences (voltage) around any closed circuit is zero.

$$\sum_{i} e_i = \sum_{i=1}^{n} I_i \cdot Ri$$

$$I_T = I_1 = I_2, V_T = V_1 + V_2, R_T = R_1 + R_2$$

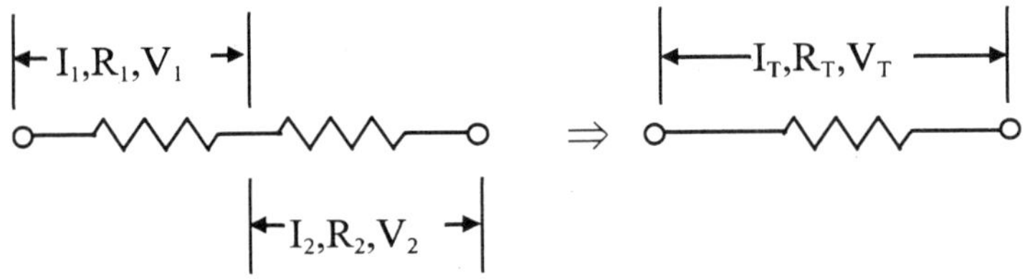

Figure 2-3 Resistors in serial connection

As in the above figure, if we make a serial connection of resistor $R_1 R_2 \cdots R_n$, and connect voltage at both ends, the current I is the same at each resistor. Therefore, the voltage drop at both ends of each resistor is calculated as the following:

$$V_1 = IR_1,\ V_2 = IR_2$$

$$\therefore V = V_1 + V_2 + \ldots V_n = (R_1 + R_2 + \cdots R_n)I$$

Therefore, the combined resistor is $R = R_1 + R_2 + \ldots R_n$, and the voltage at each resistor is

$$V_1 = \frac{R_1}{R} V\ ,\ V_2 = \frac{R_2}{R} V,\ \ldots\ldots$$

The voltage at each resistor is proportional to its own resistance.

$$I_T = I_1 = I_2,\ V_T = V_1 + V_2,\ R_T = R_1 \cdot R_2$$

As illustrated in the following figure, in serial connection of resistor $R_1 R_2 \cdots R_n$, the voltage applied to each resistor is equal.

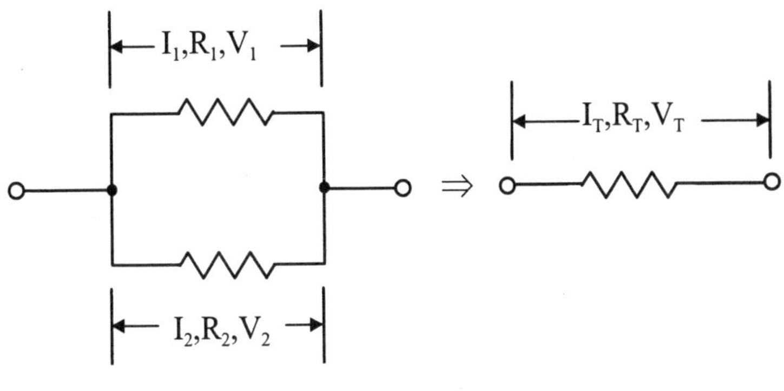

Figure 2-4 Resistors in parallel connection

$$I_1 = \frac{V}{R_1},\ I_2 = \frac{V}{R_2},\ \ldots\ldots$$

The total current I is the sum of the voltages at the resistors.

$$I = I_1 + I_2 + I_3 + \cdots + I_n = \left(\frac{1}{R_1} + \frac{1}{R_2} + \cdots + \frac{1}{R_n}\right) \cdot V = \frac{V}{R}$$

Therefore, the combined resistor is $\frac{1}{R} = \frac{1}{R_1} + \frac{1}{R_2} + \cdots + \frac{1}{R_n}$, and the voltage at each resistor is

$$I_1 = \frac{R}{R_1} \cdot I,\ I_2 = \frac{R}{R_2} \cdot I,\ \ldots\ldots$$

The voltage at each resistor is inverse proportional to its own resistance.

As current flows in a conducting wire, movement of electric charge occurs. Electric charge is moved by the external electric field from the point with higher potential energy to the point with lower potential energy. Electric resistance obstructs movement of electric charge, and friction heat is generated between the moving electric charge and the obstructing resistance. This is because the electric field is working, and the energy from the electric field becomes the heat in the conductor.

When the current I flows in the conductor with resistance R, the potential difference (voltage or electromotive force) is $V=IR$. The current of I[A] flowing in the conductor means that the electric charge of 1[C] moves for 1[sec]. Therefore, the electric field works for $V\cdot I$ [joule/sec]. This is called power, and the unit of power is Watt[W] in the MKS unit system.

$$P=V\cdot I=I^2R\,[W]$$

The heat generated by the potential difference and the current at a resistor is called Joule's heat.

The unit for heat is calorie [cal]. It is the heat required to increase the temperature of 1cc (1mℓ=1g) of pure water from 4℃ to 5℃.

One Cal is required to increase the temperature of 1 liter of pure water by 1℃, and is equivalent to 4.185 Joule.

$$\text{Calorie}=24^2Rt[Cal]=\frac{1}{4.185}{}^2Rt$$

3. Joule's heat

Joule's heat is the heat generated due to collision between electrons or with atoms, increasing the temperature of a metal. In other words, the energy given to electrons by the electric field is given to atoms due to collision with the atoms, and the thermal vibration of electrons become violent causing the temperature to increase.

Electric current flows in a conductor with resistance. Electric charge flows from the point with higher electric potential to the point with lower electric potential. The electric field (space of force) is consumed for work. Energy of the electric field is reduced. The reduced energy is converted into heat inside the conductor.

4. Power

Electric energy per unit of time

- Power $P = V \cdot I = I^2 \cdot R = \frac{V^2}{R}[W]$
- Energy $W = P \cdot t = V \cdot I \cdot t = I^2 \cdot R \cdot t[W \cdot s]$

5. Analysis of capacitor

If the numbers of neighboring conductors make a conductor system, the electric fields caused by electric charges given to each conductor influence each other. Therefore, surface charge and electric potential of each conductor must be considered for the entire conductor system.

In the following figure, 2 conductors are closed to each other in the space with the permittivity ϵ, and each conductor is connected with a conducting wire and is given the potential difference $V_{AB}[V]$. If the electric charge of $+Q[C]$ is supplied to a conductor, $-Q[C]$ is induced at the surface of the other conductor.

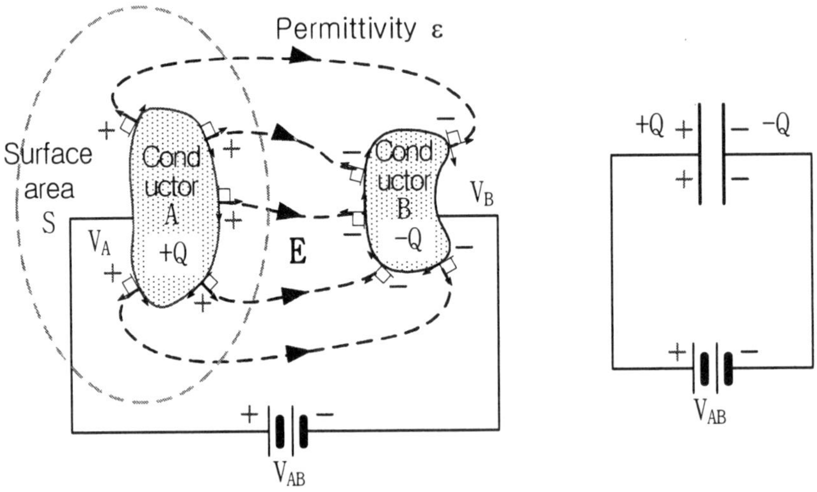

Figure 2-5 Capacity with 2 conductors

In the electromagnetic field, a conductor has equipotential, and DC invokes movement of electric charges. Electric charges distributed in a conductor exist on the surface only so that the electric field inside the conductor should become 0. When the line of electric force and dielectric flux are perpendicular to the surface of the conductor, and the potential difference between conductors according to distribution of electric charges $\pm Q[C]$ is $V_{AB}[V]$, there is a linear relationship between the distributed electric charges and the potential difference:

$$Q = C \cdot V_{AB} \Rightarrow C[F] = \frac{Q}{V_{AB}}$$

The ratio of total electric charge to the potential difference Q/V_{AB} is constant. The proportional factor C is called the capacitance of the conductor system. The unit used is coulomb[C] for electric charge, volt[V] for potential difference and Farad[F] for capacity.

Where the magnitude of electric field between conductors is E[V/m] and the surface area of the volume of the conductor with the positive electric charge +Q[C] is S[m^2], the capacity is redefined as the following according to the Gauss's law$\left(Q = \int_S D \cdot dS\right)$ and the formula of electric potential $\left(V_{AB} = -\int_B^A E \cdot dl\right)$:

$$C = \frac{Q}{V_{AB}} = \frac{\oint_S D \cdot dS}{-\int_B^A E \cdot dl} = -\epsilon \frac{\oint_S E \cdot dS}{\int_B^A E \cdot dl} [F]$$

The capacity is determined by the geometric structure of the conductor system and the permittivity of the neighboring dielectric substance, and is irrelevant to electric charges and electric potential. It is used to express the level of capacity of a conductor system to store electric charges.

A device which is composed of 2 conductors, divided by free space or dielectric substance and each connected with a conducting wire, and stores electric charges and electric energy with the electric potential applied or electric charges supplied is called a capacitor or condenser. This device is actually used in many electric circuits.

In order that a capacitor or condenser should store a large quantity of electric charges with enlarged capacity, it is required to narrow the gap between conductors, and to insert into the space a dielectric substance with high permittivity and appropriate dielectric strength (working voltage), such as paper, air, oil, mica or ceramic.

Capacitors can be connected with each other in series or in parallel.

(1) Serial connection

As illustrated in the following figure, a number of capacitors are connected with each other in series. With the quantity of electric charge applied to the capacitors at both ends, different types of electric charges are stored in equal quantity in the inner capacitors due to induction. Therefore, each capacity in a serial connection has the same quantity of electric charge. Each capacity has different capacity value and varied potential difference. The total potential difference equals the sum of potentials in each capacitor.

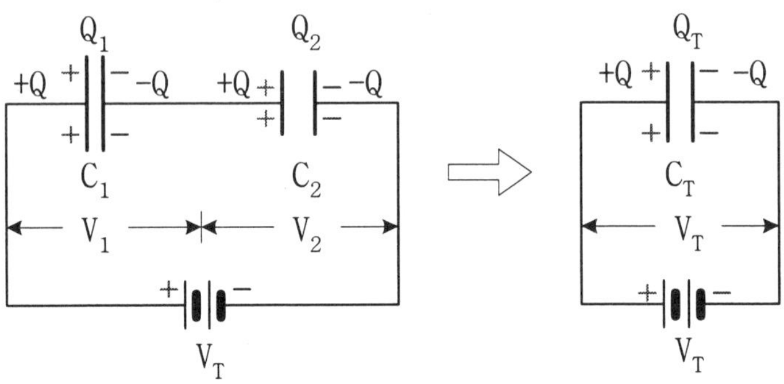

Figure 2-6 2 capacitors in serial connection

$$Q_T = Q_1 = Q_2, V_T = V_1 + V_2$$

$$\Rightarrow C_T = \frac{Q_T}{V_T} = \frac{Q_T}{V_1 + V_2} = \frac{1}{V_1/Q_T + V_2/Q_T}$$

$$= \frac{1}{1/C_1 + 1/C_2} = \frac{C_1 \cdot C_2}{C_1 + C_2}$$

$$\therefore \frac{1}{C_T} = \sum_{i=1}^{n} \frac{1}{C_i}$$

The more the capacitors in a serial connection, the more the total capacity value decreases.

(2) Parallel connection

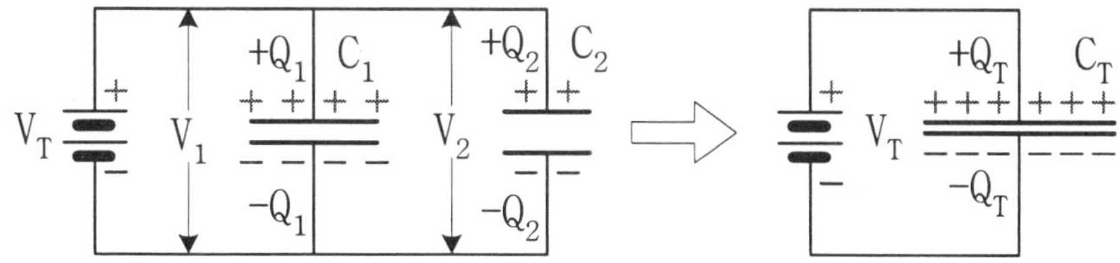

Figure 2-7 2 capacitors in parallel connection

As illustrated in the above figure, numbers of capacitors are connected in parallel to a single power supply unit. Each capacitor connected has equal potential difference. Each capacity has different capacity value, however, and hence stores varied quantities of electric charge. The total quantity of electric charge equals the sum of electric charges stored in each capacitor.

$$Q_T = Q_1 + Q_2, V_T = V_1 = V_2$$

$$\Rightarrow C_T = \frac{Q_T}{V_T} = \frac{Q_1 + Q_2}{V_T}$$

$$= \frac{Q_1}{V_T} + \frac{Q_2}{V_T} = C_1 + C_2 \therefore C_T = \sum_{i=1}^{n} C_i$$

In parallel connection of capacitors, the overall capacity value equals to the sum of the capacity value of each capacity.

6) Analysis of inductor

In a coiled electric circuit, the number of interlinked magnetic fluxes ϕ [Wb] is proportional to the voltage I[A] of the coil ($\phi = LI$[Wb]), to the number of wires N if the current is the same. And the overall interlinked magnetic flux Ψ [Wb] is also proportional ($\Psi = N\phi = LI$[Wb]).

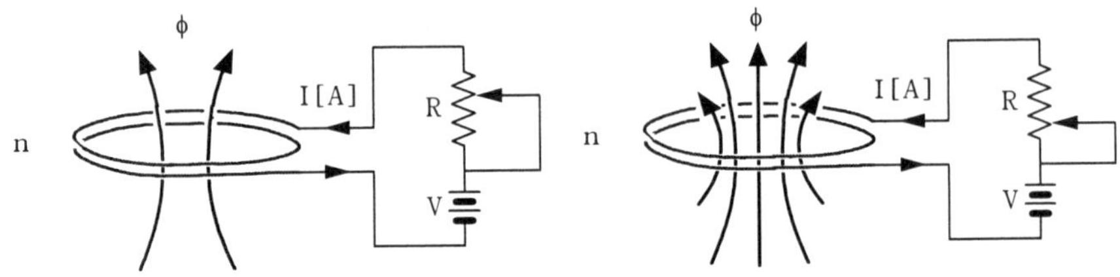

Figure 2-8 Change of interlinked magnetic flux according to the change of current

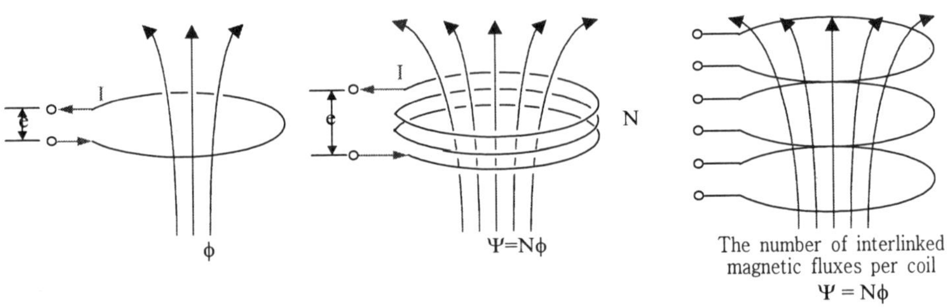

Figure 2-9 Change of the number of interlinked magnetic fluxes according to the change in the number of wires

The proportional factor L depends on the size, shape and surrounding media (magnetic permeability of magnetic substance) of the circuit. It is called a self inductance, and is defined as follows:

$$L=\frac{N\phi}{I},\ L=\frac{\phi}{I}(N=I)$$

In other words, self inductance is determined by the size, shape and surrounding media of the circuit, regardless of the intensity of the electric current, and by the number of magnetic fluxes interlinked with the coil when the unit current (1[A]) flows over the circuit.

When a unit current flows over the circuit, the number of magnetic fluxes interlinked with the coil becomes the magnitude of the self inductance. Therefore, when the number of magnetic fluxes interlinked with the circuit current 1[A] is 1[Wb], the self inductance is 1[H](Henry).

$$1[\mathrm{H}]=\frac{1[\mathrm{Wb}]}{1[\mathrm{A}]}$$

If the current of the circuit changes by time, the interlinked magnetic flux will also change. Therefore, the electromagnetic induction generates a counter electromotive force (emf) that opposes this change in current. This phenomenon is called self induction.

$$e=-\frac{d\phi}{dt}=-L\frac{dI}{dt}[\mathrm{V}]$$

Therefore, self inductance can be defined as the electromotive force induced when the circuit current changes at the rate of a unit current (1[A]) per a unit time (1[sec]).

$$1[\mathrm{H}]=\frac{1[\mathrm{V}]\cdot[\mathrm{s}]}{1[\mathrm{A}]}=1[\Omega]\cdot[\mathrm{s}]$$

The unit of magnetic permeability is the magnitude of the magnetic field H. The magnetic flux density is determined as follows with the relation of B:

$$B=\frac{\phi}{s}[\mathrm{Wb/m^2}],\ H=\frac{F}{m}\left[\frac{\mathrm{N}}{\mathrm{Wb}}=\frac{\mathrm{A}}{\mathrm{m}}\right]$$

$$\therefore \mu=\frac{B}{H}\left[\frac{\mathrm{Wb/m^2}}{\mathrm{A/m}}=\mathrm{H/m}\right]$$

In a toroid, where the self inductance is L [H], and with the instantaneous current that occurs when the switch is on, the counter emf induced to this circuit is calculated with the following formula:

$$-N\frac{d\phi}{dt}=-L\frac{di}{dt}[\mathrm{V}]$$

If the Kirchhoff's second law is applied to this circuit, the following formula is made:

$$U-L\frac{di}{dt}=Ri\,[\mathrm{V}]$$

Multiply both sides of this formula by idt, and the following formula is made:

$$Uidt=Lidi+Ri^2dt\,[\mathrm{J}]$$

In this case, the electronic energy stored in the toroid is $Lidi$ [J]. When the current is increased from 0 to I [A], the total magnetic energy is as follows:

$$W=\int_0^I Lidi=\frac{1}{2}LI^2$$

The total interlinked magnetic flux in the toroid is $N\phi$ [Wb]. Therefore, the magnetic energy is calculated as follows:

$$W=\frac{1}{2}N\phi I[\mathrm{J}]$$

$$=\frac{1}{2}\phi I[\mathrm{J}];\mathrm{N}=1$$

The formula $e=-\frac{d\phi}{dt}=-L\frac{dI}{dt}[\mathrm{V}]$ is true for a current circuit. If a number of current circuits are adjacent, the magnetic energy of the entire current system is as follows:

$$W=\frac{1}{2}L_1I_1^2+\frac{1}{2}L_2I_2^{2+}\frac{1}{2}L_3I_3^2=\frac{1}{2}\sum LI^2[\mathrm{J}]$$

$$W=\frac{I}{2}\phi_1I_1+\frac{1}{2}\phi_2I_2+\frac{1}{2}\phi_3I_3=\frac{1}{2}\sum\phi I[\mathrm{J}]$$

(1) Serial connection

Total current is the current of each coil that total induced electromotive force is the sum of induced electromotive forces of all coils.

$$I_t = I_1 = I_2,\ E_t = e_1 + e_2,\ L_T = L_1 + L_2$$

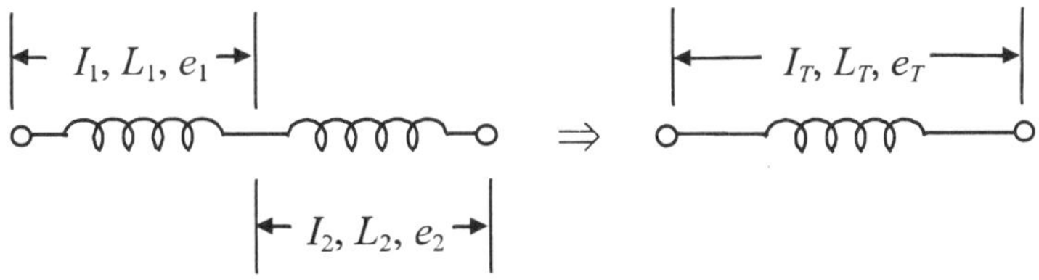

Figure 2-10 Coils in serial connection

(2) Parallel connection

Total current is sum of the currents of all coils that total induced electromotive force which is the induced electromotive force of each coil.

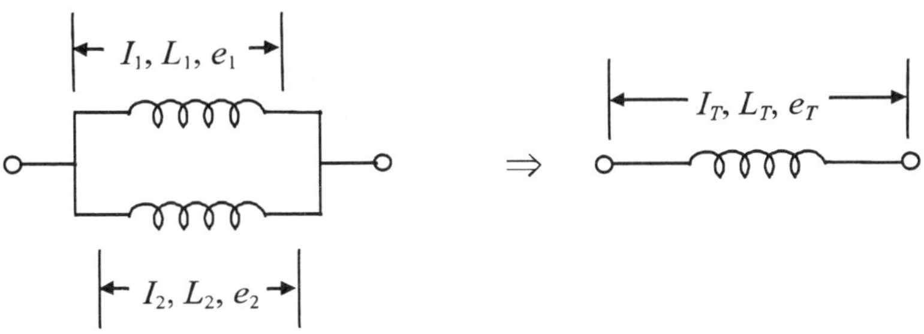

Figure 2-11 Coils in parallel connection

$$I_T = I_1 + I_2,\ e_T = e_1 = e_2,\ L_T = L_1 \cdot L_2 / (L_1 + L_2)$$

Exercise 2-1

In the following figure, calculate the current in the capacitor when the switch is on at $t=0$.

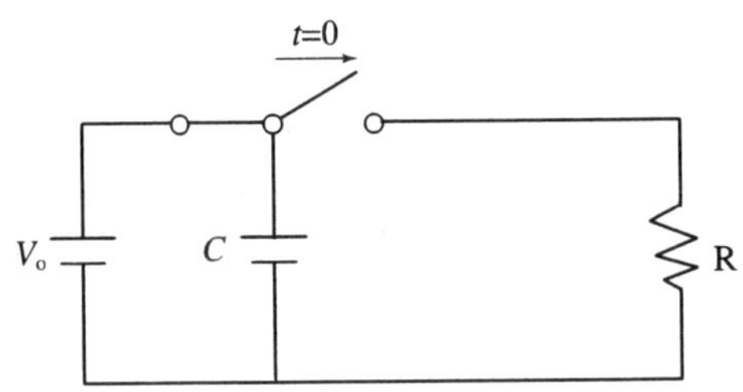

Solution

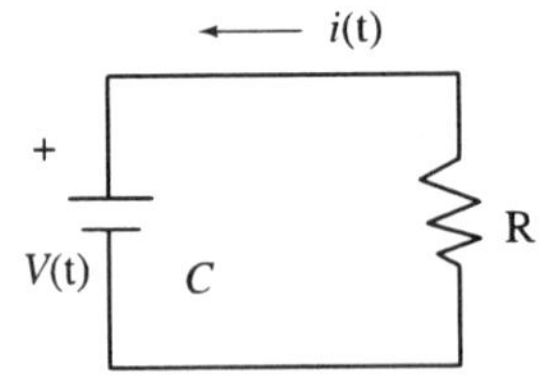

$$V(o) = V_o$$

$$V(t) + iR = 0$$

$$V(t) + R \cdot C\frac{dv(t)}{dt} = 0$$

The value of the linear differential equation $v(t) = k \cdot e^{st}$

$$ke^{st} + Rcke^{st} \cdot s = 0$$

$$ke^{st}(1 + RCS) = 0$$

$$k = 0 \text{ or } s = \frac{1}{RC}$$

$$V(t) = ke^{-\frac{1}{RC} \cdot t}$$

$$V(o) = 0 \quad k = V_O$$

$$\therefore V(t) = V_o e^{-\frac{1}{RC} \cdot t} (t \geq 0)$$

Exercise 2-2

In the following figure, calculate the current $i(t)$ when $V(o) = V_o$.

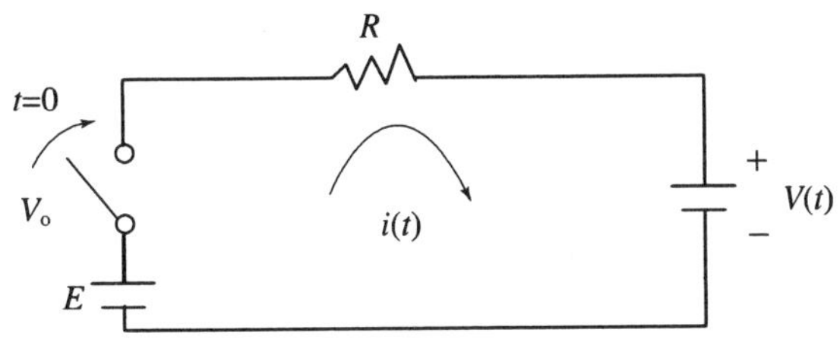

Solution

$$R \cdot C\frac{dV(t)}{dt} + V(t) = E$$

① Natural response

$$v(+) = ke^{-\frac{1}{\tau}\cdot t} \quad \tau = RC$$

② Forced response

$$t \to \infty \quad \frac{dv(t)}{dt} = 0$$

$$\therefore V = E$$

$$V(t) = E + ke^{-\frac{1}{RC}\cdot t}$$

$$V(o) = V_o,\ V_o = E + k$$

$$K = V_o - E$$

$$\therefore V(t) = E + (V_o - E)e^{-\frac{1}{RC}\cdot t} \quad (t \geq 0)$$

$$j = C\frac{dv(t)}{dt} = C \cdot \left(-\frac{1}{RC}(V_o - E)e^{-\frac{1}{RC}\cdot t}\right)$$

$$j(t) = \frac{E - V_o}{R}e^{-\frac{1}{RC}\cdot t} \quad (t \geq 0)$$

Exercise 2-3

In the following figure, calculate the current $i(t)$ when $i(o) = I_o$.

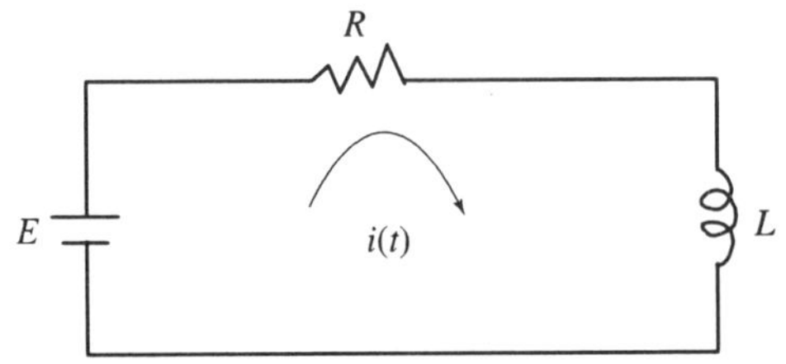

Solution

$$Ri(t) + L \cdot \frac{di(t)}{dt} = E$$

① Natural response

$$\text{in}(t) = ke^{-1/\tau \cdot t} \tau = \frac{L}{R}$$

② forced response

$$t \rightarrow 0 \quad \frac{di(t)}{dt} = 0 \,(i = \text{constant})$$

$$i_f(t) = E/R$$

$$i(t) = E/R + ke^{-\frac{1}{\tau} \cdot t}$$

$$i(o) = I_o K = I_o - E/R$$

$$\therefore i(t) = E/R + (I_o - E/R)e^{-1/\tau \cdot t}$$

Exercise 2-4

In the following figure, calculate the current $i(t)$.

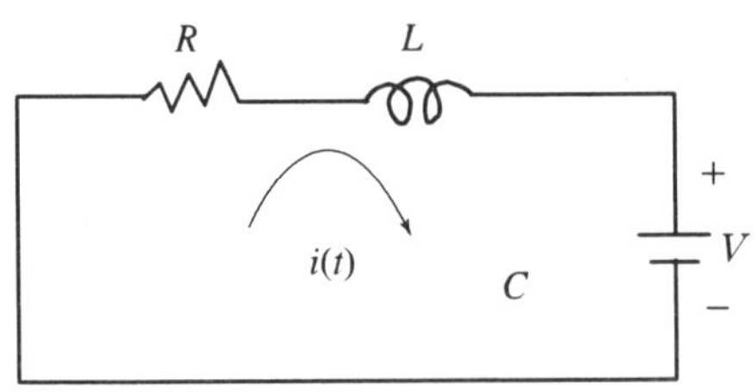

Solution

$$L\frac{di(t)}{dt}+i(t)R+V(t)=0$$

$$i(t)=C\frac{dv(t)}{dt}$$

$$LC=\frac{dv^2(t)}{dt^2}+RC\frac{dV(t)}{dt}+V(t)=0$$

$$\frac{d^2V(t)}{dt^2}+\frac{R}{L}\frac{dv(t)}{dt}+\frac{1}{LC}V(t)=0$$

$$V=ke^{st}$$

$$ke^{st}\cdot s^2+\frac{R}{C}ke^{st}\cdot s+\frac{1}{LC}ke^{st}=0$$

$$ke^{st}(s^2+\frac{R}{C}s+\frac{1}{LC})=0$$

$$K=0 \text{ or } S^2+\frac{R}{L}S+\frac{1}{LC}=0,$$

$$S_1,\ S_2=\left[-\frac{R}{L}\pm\sqrt{\left(\frac{R}{L}\right)^2-4\frac{1}{LC}}\right]/2$$

$$=-\frac{R}{2L}\pm\sqrt{\left(\frac{R}{2L}\right)^2-\left(\frac{1}{\sqrt{LC}}\right)^2}$$

$$R/2L=\alpha,\ \frac{1}{\sqrt{2C}}=w_o(\text{resonant frequency})$$

$$S_1,S_2=-\alpha\pm\sqrt{\alpha^2-w_o{}^2}$$

$$V(t)=k_e^{st}+k_2e^{s_{2t}}$$

Exercise 2-5

In the following figure, calculate the current $i(t)$.

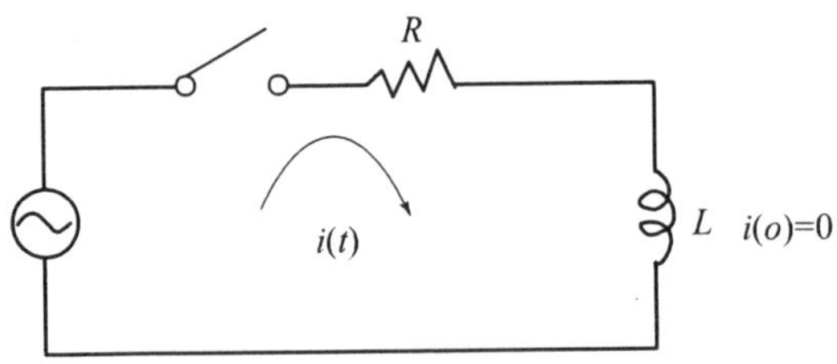

Solution

Circuit equation $i(t)R+L\frac{di(t)}{dt}=V_m\sin\omega t$

① Natural response

$$\frac{L}{R}\frac{di(t)}{dt}+i(t)=0$$

$$i_n(t)=ke^{-\frac{1}{\tau}t} \quad \tau=\frac{L}{R}(t\geq 0)$$

② Forced response

Exponential form & polar form

$$i(t)=e^{j\omega t}=\cos\omega t+j\sin\omega t$$

$$V_{L(t)}=L\cdot j\omega e^{j\omega t}=j\omega L\cdot i(t)$$

Therefore, $j\omega L$ is the component of the impedance.

$$V_c(t)=\frac{1}{C}\cdot\frac{1}{j\omega}e^{j\omega t}=\frac{1}{j\omega C}\cdot i(t)$$

The component of the impedance is $\frac{1}{j\omega C}$.

The polar form of $A<\theta$ equals the exponential form of $Ae^{j\omega t+\theta}$.

Supposing the sinusoidal source as the exponential source, the value can be found by taking the real part ($Re\{e^{j\omega t}\}$) or an imaginary part ($I_m\{e^{j\omega t}\}$) of the calculation result. Therefore,

$$\text{I} = \frac{V}{Z} = \frac{V_m \angle 0}{R + j\omega L} = \frac{V_m \angle 0}{\sqrt{R^2 + \omega^2 L^2} \angle \tan^{-1} \omega L / R}$$

$$Z_m = \sqrt{R^2 + \omega^2 L^2}$$

$$\angle \theta = \tan^{-1} \omega L / R$$

$$i_f(t) = \frac{V_m}{Z_m} sin(\omega t - \theta)$$

$$i(t) = i_f(t) + ke^{-\frac{1}{\tau}t} \quad i(o) = 0$$

$$k = 0 - \frac{V_m}{Z_m} sin(-\theta) = \frac{V_m}{Z_m} sin\theta$$

$$i(t) = \frac{V_m}{Z_m} \left[\sin(\omega t - \theta + \sin\theta e^{\frac{-1}{\tau}t} \right] (t \geqq 0)$$

Another way to calculate the impedance from the sinusoidal source is below.

$$i(t) = \sin\omega t$$

$$V_L(t) = L \cdot \cos\omega t \cdot \omega = \omega L \sin(\omega t + 90°)$$

$$= j\omega L \sin\omega t = j\omega L \cdot i(t)$$

$$V_c(t) = \frac{1}{C} \int i(t) dt = -\frac{1}{\omega C} cos\omega t$$

$$= -\frac{1}{\omega C} sin(\omega t + 90°) = -\frac{j}{\omega C} sin\omega t$$

$$= \frac{1}{j\omega C} \cdot i(t)$$

Exercises

(1) In the following figure, calculate the current $i(t)$.

$V_g(t) = 10\sin 10t \quad V_c(0) = 0$

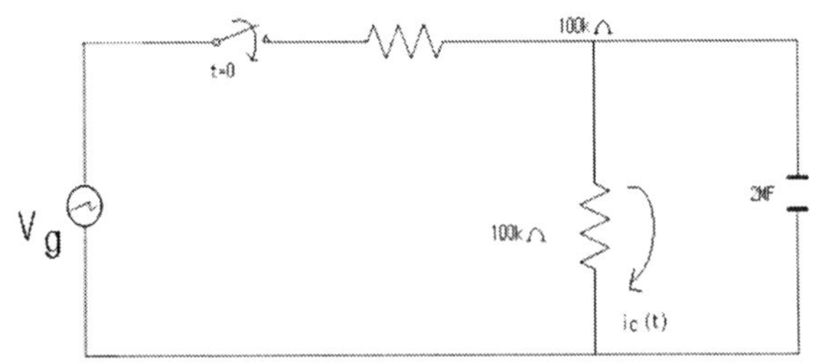

(2) In the following figure, calculate the current $i(t)$. $V_g(t) = 6\,VDC$

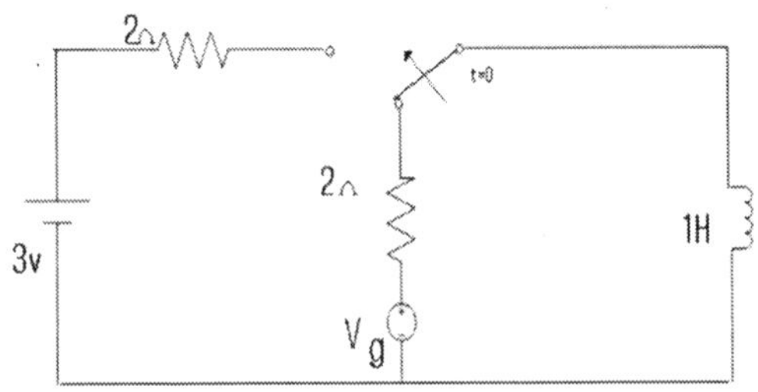

(3) In the following figure, calculate the current $i(t)$. $I(0) = 0 \cdot e(t) = u(t)$.

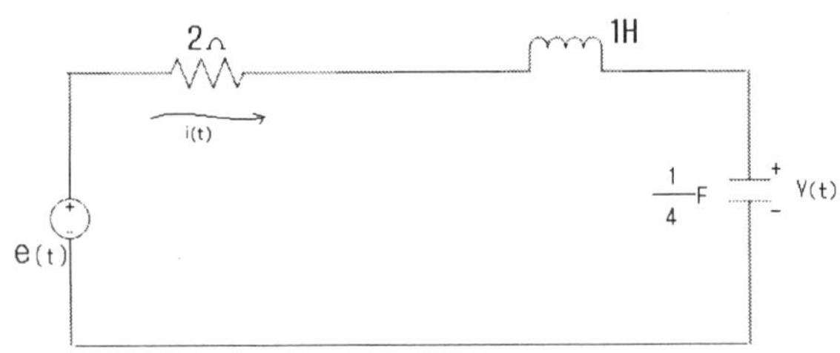

Answers

(1) Theverin Theorem

$$R=\frac{100K\times 100K}{100K+100K}=50K\Omega$$

$$V_g{}'=\frac{V_g}{200K\Omega}\cdot 100K\Omega$$

$$=\frac{V_g}{2}5\sin 10t$$

$$\tau=RC=50k\Omega\times 2\mu F=0.1$$

$$\frac{1}{j\omega c}=\frac{1}{j\times 10\times 2\mu F}$$

$$=-j50k$$

Using phasor

$$i_n(t)=ke^{-10t}$$

$$i=\frac{V_g{}'}{R+\frac{1}{j\omega c}}=\frac{5\angle 0°}{50k-j50k}$$

$$=\frac{5\angle 0°}{50k\sqrt{2}\angle -45°}=\frac{1}{10k\sqrt{2}}\angle 45°$$

$$i_f=\frac{1}{10k\sqrt{2}}sin(10t+45°)mA$$

$$i(t)=ke^{-10t}+\frac{1}{10k\sqrt{2}}sin(10t+45°)mA$$

$$i(0)=0=k+\frac{1}{10k\sqrt{2}}\cdot\frac{1}{\sqrt{2}}=k+\frac{1}{20}$$

$$k=-\frac{1}{20}$$

$$i(t)=-\frac{1}{20}e^{-10t}+-\frac{1}{10\sqrt{2}}sin(10t+45°)mA$$

(2) $t < 0$

$$i(0) = \frac{6}{2} = 3.A$$

$t > 0$

$$\tau = \frac{L}{R} = \frac{1}{2}$$

$$i_n(t) = ke^{-2t}$$

$$i_f(t) = \frac{3}{2} = 1.5$$

$$i(t) = 1.5 + ke^{-2t}$$

$$i(0) = 3 = 1.5 + k$$

$$k = 1.5$$

$$i_L(t) = 1.5 + 1.5e^{-2t}$$

(3) $S^2 + \frac{K}{2}S + \frac{1}{LC} = 0$

$$S^2 + 2S + 4 = 0$$

$$S^2 \frac{-2 \pm \sqrt{4-16}}{2} + \frac{-2 \pm \sqrt{-12}}{2}$$

$$S = -1 \pm j\sqrt{3}$$

$$V_n(t) = k_1 e^{(-1+j\sqrt{3})t} + k_2 e^{(-1-j\sqrt{3})t}$$

$$\nu_f = 1 \cdot (t - > \infty)$$

$$V(t) = 1 + k_1 e^{(-1+j\sqrt{3})t} + k_2 e^{(-1-j\sqrt{3})t}$$

$$V(0) = 1 + k_1 + k_2 = 0$$

$$k2 = -1 - k_1$$

$$i(t) = \frac{1}{4}\left[(-1) + j\sqrt{3})k_1 e^{(-1+j\sqrt{3})t} + (-1 - j\sqrt{3})k_2 e^{(-1-j\sqrt{3})t}\right]$$

$$i(0) = 0$$

$$\frac{1}{4}\left[(-1) + j\sqrt{3})k_1 + (-1 - j\sqrt{3})k_2\right] = 0$$

$$(-1) + j\sqrt{3})k_1 + (-1 - j\sqrt{3})(-1 - k_1) = 0$$

$$-k_1 + j\sqrt{3}k_1 + 1 + j\sqrt{3} + k_1 + j\sqrt{3}k_1 = 0$$

$$k_1 = \frac{-1 - j\sqrt{3}}{j2\sqrt{3}} = \frac{-\sqrt{3} + j}{2\sqrt{3}}$$

$$k_2 = \frac{-\sqrt{3} - j}{2\sqrt{3}}$$

$$i(t) = 1 + \left(\frac{-\sqrt{3} + j}{2\sqrt{3}}\right)e^{(-1+j\sqrt{3})t} + \left(\frac{-\sqrt{3} - j}{2\sqrt{3}}\right)e^{(-1-j\sqrt{3})t}$$

Chapter 3

Signal Analysis

Communication technology refers to all the technical means required to send desired signals to destinations. From a standpoint of a communication, the desired information is called a signal, and all other information is called interference or noise. The types of signals existing in the natural world are infinite, and therefore, the interpretation and expression of signals is very complicated. Now, we will find out the types of basic signals and the representations for signal processing.

The sine wave plays a very important role in the interpretation of signals, as well as in modulation and demodulation. The following figure illustrates how the sine wave is generated. When the unit vector fixed to the origin goes round a column at the interval of T_o, the perpendicular component of the vector makes a sine wave as time passes. Frequency f_o and angular frequency ω_o are as follows:

$$f_o = \frac{1}{T_o} 1/\text{sec}, \omega_o = 2\pi f_o \text{rad/sec}$$

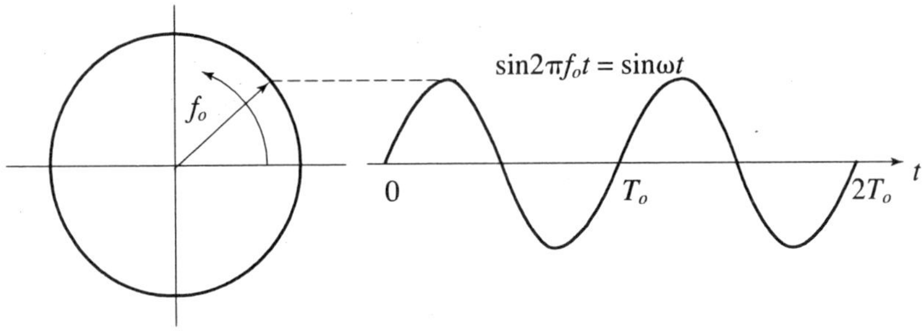

Figure 3-1 Sinusoidal signal

Exponential function is represented as $f(t) = A\exp(st)$.

Where, A and s are complex numbers in general. When s is a complex number, it is called a complex exponential. If $s = j\beta$, the following formula is made based on Euler's formula.

$$A\exp(j\beta t) = A\cos\beta t + jA\sin\beta t$$

$s = \alpha + j\beta$, and for complex numbers, the following damped oscillation occurs:

$$A\exp(\alpha + j\beta t) = A\exp(\alpha)\cos\beta t + jA\exp(\alpha)\sin\beta t$$

Exponential functions are used as supplementary values of differential equations. Therefore, they are necessary to express the transient state of the system, and play key roles in Fourier transform in expressing signals in the frequency domain.

(1) Signals for communication

A step function is defined as follows:

$$u(t)=\begin{cases}1, & t\geq 0\\ 0, & t<0\end{cases}$$

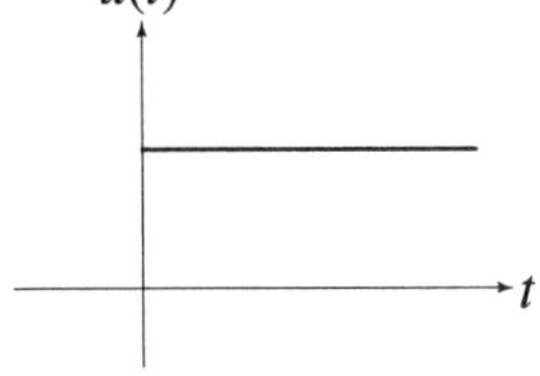

Figure 3-2 Unit step function

Using step functions, it is easy to indicate the reference clock for signal input or system operation. Impulse function can be defined as follows:

$$\delta(t)=\begin{cases}\infty, t=0\\ 0, t\neq 0\end{cases}$$

$$\int_{-\infty}^{\infty}\delta(t)dt=1$$

Impulse function is used as an important tool for the interpretation of all signals and systems.

Multiply a function $f(t)$ by $\delta(t)$, and $f(t)\delta(t)=f(0)\delta(t)$.

$$\int_{-\infty}^{t}\delta(t)dt=\begin{cases}0, t<0\\ 1, t\geq 0\end{cases}=u(t)$$

Therefore, the following formula is made:

$$\int_{-\infty}^{\infty}f(t)\delta(t-T)dt=\int_{-\infty}^{\infty}f(T)\delta(t-T)dt$$

$$=f(T)\int_{-\infty}^{\infty}\delta(t-T)dt$$

$$=f(T)$$

The above formula can be extended to a more universal form such as the following:

$$f(t)=\int_{-\infty}^{\infty} f(\tau)\delta(t-\tau)d\tau$$

Fourier series is the way to express a periodic function as a linear combination of sine waves in harmonic relation. Non-sinusoidal function $f(t)$ with the interval T can be represented in the following formula:

$$f(t)=a_0+\sum_{n=1}^{\infty}[a_n \cos n\omega_o t+b_n \sin n\omega_o t] \quad \left(\omega_o=\frac{2\pi}{T}\right)$$

The above expression must satisfy the condition that $f(t)$ is a periodic function. Using the fact that harmonics use the fundamental period as the common period, the following formula is made:

$$f(t+T)=a_0+\sum_{n=1}^{\infty}[a_n \cos n\omega_o(t+T)+b_n \sin n\omega_o(t+T)]$$

$$=a_0+\sum_{n=1}^{\infty}[a_n \cos n\omega_o t+b_n \sin n\omega_o t]$$

$$=f(t)$$

Coefficients a_n and b_n can be calculated with orthogonality between harmonics. Multiply both sides by a harmonics $\cos k\omega_o t$. Integrate it for a period T, and the following result is acquired.

$$\int_{t_o}^{t_o+T} f(t)\cos k\omega_o t\,dt$$

$$=\int_{t_o}^{t_o+T}\left[a_0+\sum_{n=1}^{\infty}[a_n \cos n\omega_o t+b_n \sin n\omega_o t]\right]\cos k\omega_o t\,dt$$

$$=a_o\int_{t_o}^{t_o+T}\cos k\omega_o t\,dt+\sum_{n=1}^{\infty}a_n\left[\int_{t_o}^{t_o+T}\cos n\omega_o t\cos k\omega_o t\right]dt$$

$$\sum_{n=1}^{\infty}b_n\left[\int_{t_o}^{t_o+T}\sin n\omega_o t\cos k\omega_o t\right]dt$$

$$= Ta_k/2$$

Therefore, coefficient a_k of the harmonics $\cos k\omega_o t$ is represented as follows:

$$a_k = \frac{2}{T}\int_{t_o}^{t_o+T} f(t)\cos k\omega_o t\,dt$$

In the same way, to find the coefficient b_k, multiply both sides by $\sin k\omega_o t$ and integrate it for a period.

$$\int_{t_o}^{t_o+T} f(t)\sin k\omega_o t\,dt$$

$$= \int_{t_o}^{t_o+T}\left[a_0 + \sum_{n=1}^{\infty}[a_n\cos n\omega_o t + b_n\sin n\omega_o t]\right]\sin k\omega_o t\,dt$$

$$= a_0\int_{t_o}^{t_o+T}\sin k\omega_o t\,dt + \sum_{n=1}^{\infty} a_n\left[\int_{t_o}^{t_o+T}\cos n\omega_o t\sin k\omega_o t\right]dt$$

$$\sum_{n=1}^{\infty} b_n\left[\int_{t_o}^{t_o+T}\sin n\omega_o t\sin k\omega_o t\right]dt$$

$$= Tb_k/2$$

Therefore, the coefficient b_k of the harmonics $\sin k\omega_o t$ is represented as follows:

$$b_k = \frac{2}{T}\int_{t_o}^{t_o+T} f(t)\sin k\omega_o t\,dt$$

The coefficient a_0 is the average of $f(t)$ for a period.

$$a_0 = \frac{1}{T}\int_{t_o}^{t_o+T} f(t)\,dt$$

Complex exponential function is represented as a Fourier series in a simple form.

$$f(t) = a_0 + \sum_{n=1}^{\infty}[a_n\cos n\omega_o t + b_n\sin n\omega_o t] \quad \left(\omega_o = \frac{2\pi}{T}\right)$$

The above formula can be represented as the following with the complex exponential function:

$$f(t)=\sum_{n=-\infty}^{\infty} c_n \exp(jn\omega_o t)$$

In the above formula, the coefficient c_k is calculated as the following:

$$c_k=\frac{1}{T}\int_{t_o}^{t_o+T} f(t)\exp(-jk\omega_o t)dt$$

This is called an exponential Fourier series, and is represented as follows:

$$f(t)=\sum_{n=-\infty}^{\infty} c_n \exp(jn\omega_o t) \quad \left(\omega_o=\frac{2\pi}{T}\right)$$

$$c_n=\frac{1}{T}\int_{t_o}^{t_o+T} f(t)\exp(-jn\omega_o t)dt$$

(2) Impulse response of system

The relation between system input and output is represented as follows:

$$y(t)=T[x(t)]$$

The operator $T[\cdot]$ is applied to the time variable t only, and converts the time function $x(t)$ to another function $y(t)$.

The signal output for the impulse function input is called the system impulse response, and is represented as follows:

$$h(t)=T[\delta(t)]$$

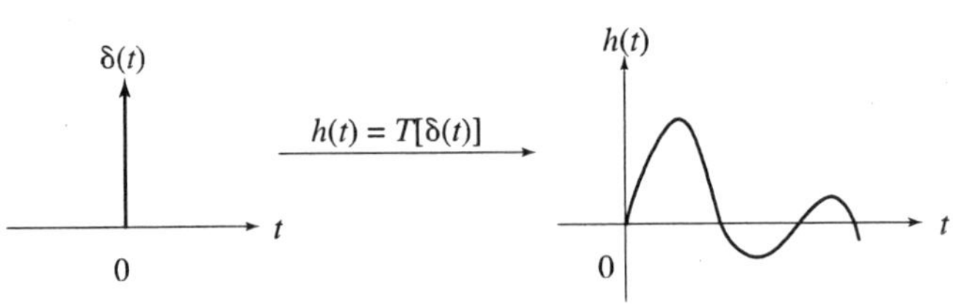

Figure 3-3 Impulse response

System input $x(t)$ is represented as the following:

$$x(t) = \int_{-\infty}^{\infty} x(\tau)\delta(t-\tau)d\tau$$

The operator $T[\cdot]$ is applied to the time variable t only. Therefore, the output $y(t)$ is represented as the following:

$$y(t) = T[x(t)]$$

$$y(t) = T[x(t)] = T\left[\int_{-\infty}^{\infty} x(\tau)\delta(t-\tau)d\tau\right]$$

$$= \int_{-\infty}^{\infty} x(\tau)\,T[\delta(t-\tau)]d\tau$$

If the system is a time-invariant system,

$h(t-\tau) = T[\delta(t-\tau)]$, and

$$y(t) = \int_{-\infty}^{\infty} x(\tau)\,T[\delta(t-\tau)]d\tau = \int_{-\infty}^{\infty} x(\tau)h(t-\tau)\,d\tau.$$

This is called convolution, and is represented as the following:

$$x(t) * h(t) \equiv \int_{-\infty}^{\infty} x(\tau)h(t-\tau)d\tau$$

Convolution defines the relation between system input and output clearly in the aspect of time.

Exercise 3-1

Calculate the output of the following system using convolution.

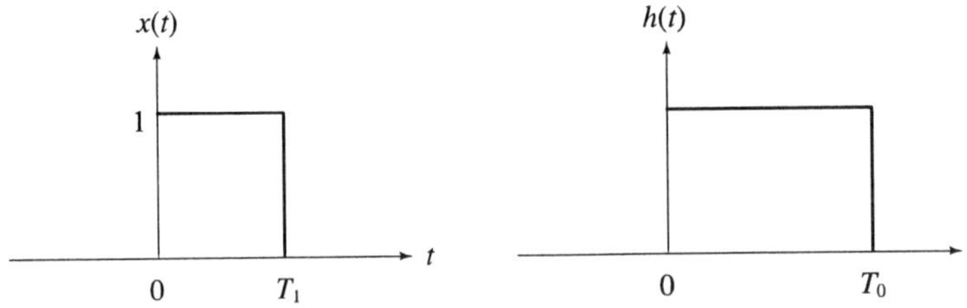

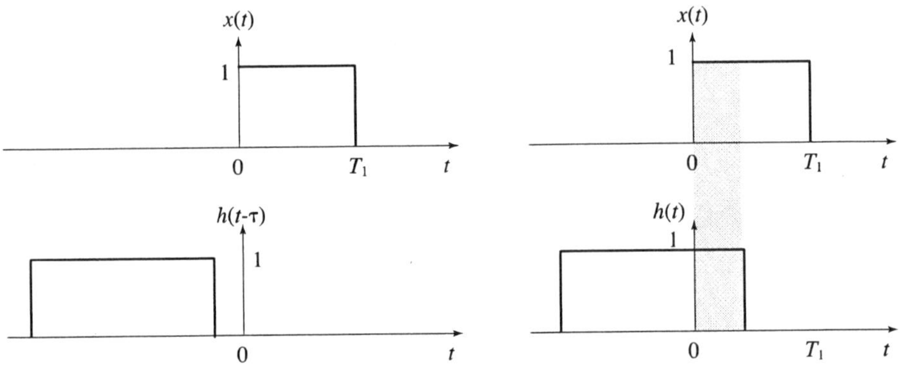

(a) $t < 0$ (b) $(0 < t < T_1)$

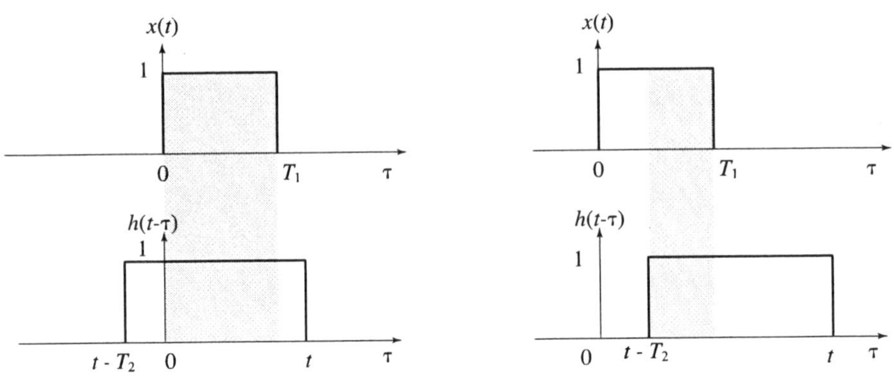

(c) $(T_1 < t < T_2)$ (d) $(T_2 \leq t < T_1 + T_2)$

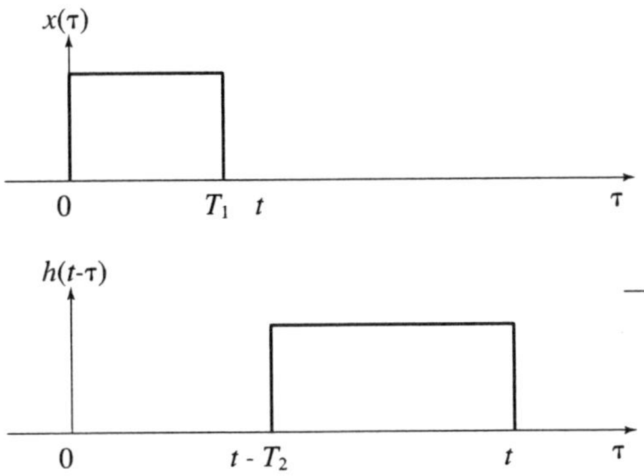

(e) $(t > T_1 + T_2)$

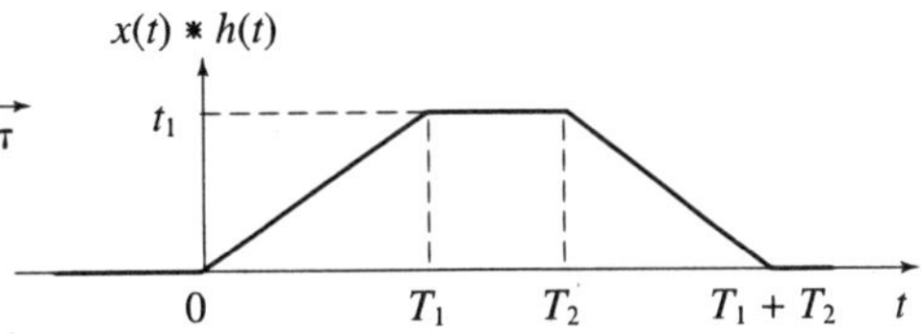

(f)

Solution

(a) $t < 0$: There is no overlapped part between $x(\tau)$ and $h(t-\tau)$, and the integral value is 0.

$$x(t)*h(t) = 0, (t < 0)$$

(b) $0 \leqq t < T_1$: The overlapped part between $x(\tau)$ and $h(t-\tau)$ is $0 \leqq \tau < t$, and hence,

$$x(t)*h(t) = \int_0^t (1)dt = t, (0 \leqq t < T_1)$$

(c) $T_1 \leqq t < T_2$: The overlapped part between $x(\tau)$ and $h(t-\tau)$ is $0 \leqq \tau < T_1$, and hence,

$$x(t)*h(t) = \int_0^{T_1} (1)dt = T_1, (T_1 \leqq t < T_2)$$

(d) $T_2 t < T_1 + T_2$: The overlapped part between $x(\tau)$ and $h(t-\tau)$ is $t - T_2 \leqq \tau < T_1$, and hence,

$$x(t)*h(t) = \int_{t-T_2}^{T_1} (1)dt = T_1 + T_2 - t, (T_2 \leqq t < T_1 + T_2)$$

(e) $t > T_1 + T_2$: There is no overlapped part between $x(\tau)$ and $h(t-\tau)$, and hence,

$$x(t)*h(t) = 0, (t > T_1 + T_2)$$

Exercise 3-2

Various automation processes could be controlled by the following computer system. The figure illustrates the process of measuring and recording the temperature of a micro oven. Output signals of an amplifier are sampled through an analog-digital converter every 10 seconds. The sampled signal is scaled by the computer that the celsius temperature of the micro oven should be divided by 10. Temperature information processed in this procedure is stored in Array X, RF noise-filtered, and the output from the filter is stored in array Y. Array Y values are displayed on the screen through the digital-analog converter by the operator's command. When sampling is started, the temperature of the micro oven is 20° C, and is increasing by 2° C per second. The initial value of the digital filter is set to 0. This system is an LTI system, and the formula of the digital filter is as follows:

$$y[n]-0.5y[n-1]=0.5x[n]$$

Calculate, from the analog signal, the input signal $x[n]$ of the digital filter, the impulse response $h[n]$, and then, outputs $y[1]$, $y[2]$ and $y[3]$ by using convolution.

Amp→ADC→Array X→Digital filter→Array Y→DAC

If we convert a signal into the frequency domain, we can get an important information called spectrum, which we could not get in the time domain.

We see spectrum of signal through Fourier transform. Fourier series is limited to periodic signal, and Fourier transform extends Fourier series and enables conversion of non-periodic signal into the frequency domain. Extension of Fourier series to Fourier transform converges the period of a periodic signal to infinity, and creates a non-periodic function.

Where a periodic function with the period T is $\widetilde{f}_T(t)$, select a function (t) for a period $-T/2 \le t \le T/2$. Apply Fourier series to the periodic function $\tilde{f}_T(t)$ as follows:

$$\tilde{f}_T(t) = \sum_{n=-\infty}^{\infty} c_n \exp(jn\omega_o t),$$

$$c_n = \frac{1}{T}\int_{-T/2}^{T/2} \tilde{f}_T(t)\exp(-jn\omega_o t)dt$$

In the above formula, take the limit as the period ($T \to \infty$). The frequency gap ω_o of the line spectrum converges to the infinitesimal, and it is indicated as $d\omega$.

$$\omega_o = \frac{2\pi}{T} \xrightarrow{T\to\infty} d\omega = \frac{2\pi}{T} \text{ or } T = \frac{2\pi}{d\omega}$$

As ω_o becomes infinitesimal, the line spectrum frequency $n\omega_o$ is represented with the continuous variable ω as shown in the following:

$$n\omega_o \xrightarrow{T\to\infty} \omega$$

The above two changes are applied to Fourier series as follows,

$$\tilde{f}_T(t) = \sum_{n=-\infty}^{\infty} c_n \exp(jn\omega_o t),$$

$$c_n = \frac{1}{T}\int_{-T/2}^{T/2} \tilde{f}_T(t)\exp(-jn\omega_o t)dt$$

and we will get the following expression.

$$\frac{c_n}{d\omega} = \frac{1}{2\pi}\int_{t=-\infty}^{\infty} \tilde{f}_T(t)\exp(-j\omega t)dt$$

This formula is substituted for

$\tilde{f}_T(t) = \sum_{n=-\infty}^{\infty} c_n \exp(jn\omega_o t)$, and using $T\to\infty$, the following change is made.

$$\tilde{f}_T(t) \xrightarrow{T\to\infty} f(t), \quad \sum_{n=-\infty}^{\infty} \xrightarrow{T\to\infty} \int_{\omega=-\infty}^{\infty}$$

By using the change, the following formula is made.

$$f(t) = \frac{1}{2\pi}\int_{-\infty}^{\infty} F(\omega)\exp(j\omega t)d\omega,$$

$$F(\omega) = \int_{-\infty}^{\infty} f(t)\exp(-j\omega t)dt.$$

The following formula shows an example of transformation of impulse function.

$$F[\delta(t)] = \int_{-\infty}^{\infty} \delta(t)\exp(-j\omega t)dt$$

$$= \exp(0)\int_{-\infty}^{\infty} \delta(t)dt$$

$$= 1$$

Therefore, the following transform pair is acquired.

$$\delta(t) \xleftrightarrow{F} 1$$

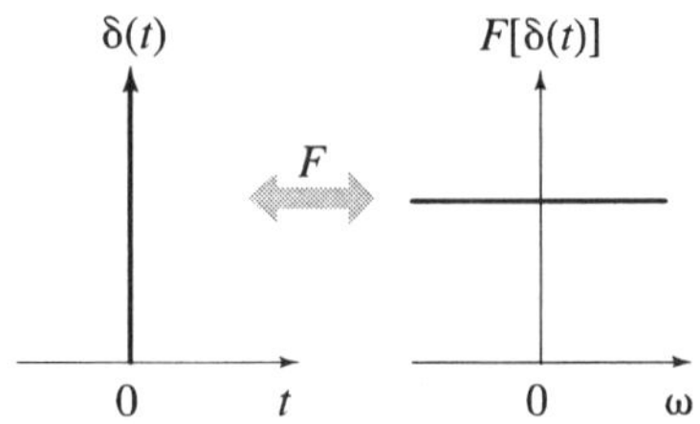

Exercise 3-3

Find the Fourier transform of exponential function $f(t) = \exp(-at)u(t)$, $a > 0$.

Solution

$$F[\exp(-at)u(t)] = \int_{-\infty}^{\infty} \exp(-at)\exp(-j\omega t)u(t)dt$$

$$= \int_{-\infty}^{\infty} \exp[-(a+j\omega)t]dt$$

$$= \frac{1}{a+j\omega}, a > 0$$

Power is a variation of energy by time, that is, a differentiation. We can get the average power by averaging the energy variation. The total energy E and the average power P of a signal $f(t)$ are defined as follows:

$$E=\int_{-\infty}^{\infty}|f(t)|^2\,dt$$

The correlation function for a signal $f(t)$ is defined as follows:

$$R_f(\tau)=\lim_{T\to\infty}\frac{1}{2T}\int_{-T}^{T}f^*(t)f(t+\tau)dt \quad \text{(Complex signal)}$$

$$=\lim_{T\to\infty}\frac{1}{2T}\int_{-T}^{T}f(t)f(t+\tau)dt \quad \text{(Real signal)}$$

The above function expresses correlation of a signal $f(t)$, and is called a autocorrelation function.

(3) Sampling theory and continuous time signal/discrete time signal

We can get a discrete time signal by multiplying a continuous time signal by an impulse train. The characteristics of the discrete time signal can be utilized in communication and signal processing. The impulse train in a time domain is expressed as follows:

$$x_s(t)=x_c(t)s(t)$$

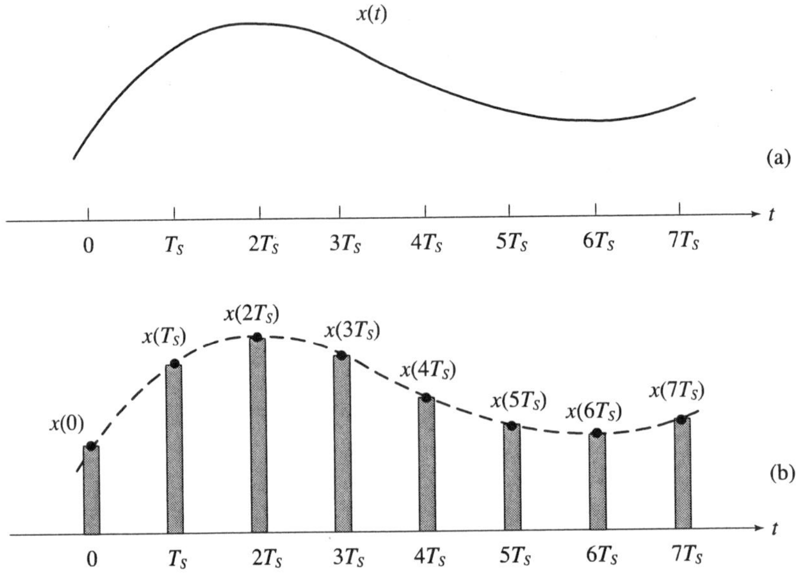

Figure 3-4 Continuous time signals

Multiplication of two continuous time signals is represented as a convolution in the frequency domain.

$$x_s(t) = x_c(t)s(t) \xleftrightarrow{CTFT} X_s(j\Omega) * S(j\Omega)$$

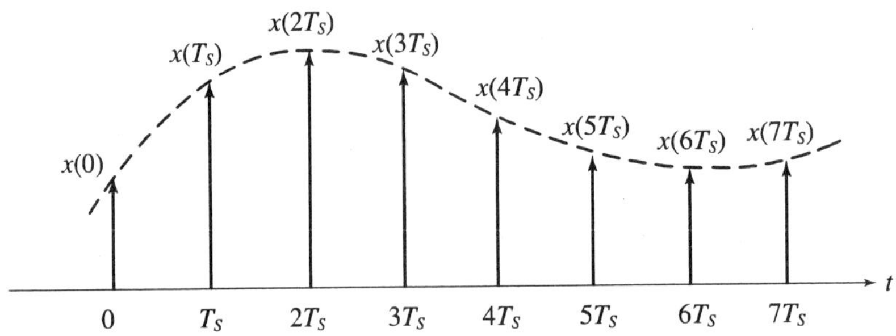

Figure 3-5 Discrete time signals

CTFT of a function $x(t)$ is represented as the following according to the definition:

$$X(j\Omega) = \int_{-\infty}^{\infty} x(t) e^{-j\Omega t} dt$$

As an inversion formula, $x(t)$ and $s(t)$ are calculated as follows according to the definition:

$$x(t) = \frac{1}{2\pi} \int_{-\infty}^{\infty} X(j\Omega) e^{-j\Omega t} dt$$

$$s(t) = \sum_{n=-\infty}^{\infty} \delta(t - nT) = \tilde{\delta}(t)$$

Fourier Transform (CTFT) is represented as the following by using Discrete Fourier Series (DFS):

$$\tilde{\delta}(t) = \sum_{n=-\infty}^{\infty} X_k e^{jk\Omega_0 t}$$

Coefficients of DFS is calculated in the following formula according to the definition:

$$X_k = \frac{1}{T} \int_{-T/2}^{T/2} \delta(t) e^{-jk\Omega_0 t} dt = \frac{1}{T}$$

Therefore, $s(t)$ is represented as the following DFS:

$$s(t) = \tilde{\delta}(t) = \sum_{n=-\infty}^{\infty} \delta(t - nT) = \frac{1}{T} \sum_{k=-\infty}^{\infty} e^{jk\Omega_0 t}$$

Calculate CTFT for the above $s(t)$ by calculating CTFT of a simple function first.

$$1 \leftrightarrow 2\pi\delta(\Omega)$$

$$e^{j\Omega_0 t} \leftrightarrow 2\pi\delta(\Omega - \Omega_{0)}$$

$$\frac{1}{T} \sum_{k=-\infty}^{\infty} e^{lk\Omega_0 t} \leftrightarrow \frac{2\pi}{T} \sum_{k=-\infty}^{\infty} \delta(\Omega - k\Omega_0) = \Omega_0 \sum_{k=-\infty}^{\infty} \delta(\Omega - k\Omega_0) = \Omega_0 \tilde{\delta}(t)(\Omega)$$

Therefore, as mentioned above, CTFT for an impulse train can be calculated with the following convolution:

$$X_s(j\Omega) = \frac{1}{2\pi} X_c(j\Omega) * \frac{2\pi}{T} \sum_{k=-\infty}^{\infty} \delta(\Omega - \Omega_0)$$

$$= \frac{1}{T} \sum_{k=-\infty}^{\infty} X_c(j(\Omega - \Omega_0))$$

In the above formula, Fourier transform of the impulse train value for a continuous time signal is a repetition of Fourier transform of the original signal. The entire gain is I/T.

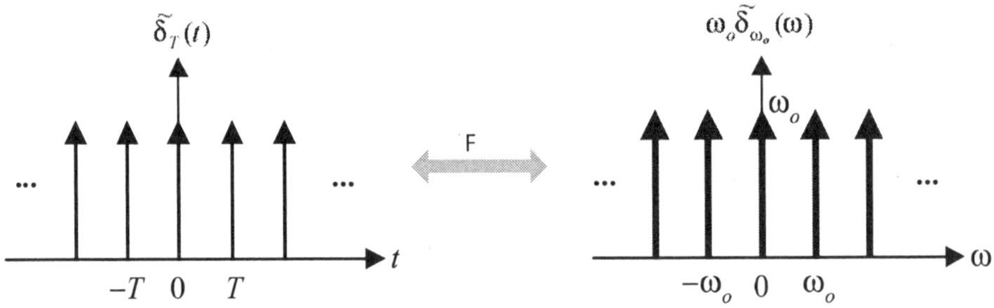

Figure 3-6 Fourier transform of Impulse train

Let's set up a relational equation of CTFT and DTFT. First, CTFT for the impulse train is as follows:

$$x_s(t) = x_c(t)s(t) = x_c(t) \sum_{k=-\infty}^{\infty} \delta(t - nT)$$

$$= \sum_{n=-\infty}^{\infty} x_c(nT)\delta(t-nT) \leftrightarrow X_s(j\Omega) = \sum_{n=-\infty}^{\infty} x_c(nT) e^{-j\Omega n T}$$

DTFT for $x[n]$ is as follows:

$$X(e^{j\omega}) = \sum_{n=-\infty}^{\infty} x[n]e^{-j\omega n} = \sum_{n=-\infty}^{\infty} x_c[nT]e^{-j\omega n}$$

$$\Rightarrow X(e^{j\Omega T}) = \sum_{n=-\infty}^{\infty} x_c[nT]e^{j\Omega n T}$$

Eventually, the following relational equation between CTFT and DTFT is deduced from the above two equations:

$$X(e^{j\omega})\mid_{\omega=\Omega T} = X_s(j\Omega)$$

The following frequency characteristics of the discrete time signal is drawn from the above result:

$$X(e^{j\omega}) = \frac{1}{T}\sum_{k=-\infty}^{\infty} X_c(j\frac{\omega - 2\pi k}{T})$$

$X(e^{j\omega})$ is $X_c(jw/T)$ signal repeated at an interval of 2π. Of course, the gain is $1/T$.

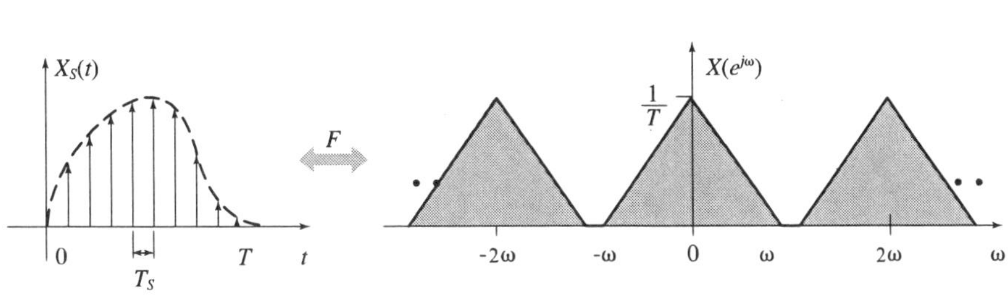

Figure 3-7 Fourier transform of signal

As we got a discrete time signal from a continuous time signal by using the impulse train, we found that Fourier transform of the original is repeated in the frequency domain. Therefore, if Fourier transform of a continuous time signal contains a RF component (that is, if the tail is long due to an infinitely large frequency component), when acquiring a discrete signal, it is inevitable that the tail components overlapp in the repeated part of the frequency domain. This phenomenon is called aliasing. We need to be careful to avoid aliasing when acquiring a discrete time signal.

When acquiring a discrete time signal from a continuous time signal through sampling, in order to avoid aliasing, we should select appropriate sampling time T. The shorter the T, the farther the $2\pi/T$ gets from the origin, and hence, aliasing gets smaller. If T is too short, however, it is ineffective because the data sampled per unit time becomes too much.

$$\Omega_s = \frac{2\pi}{T} = 2\Omega_N$$

where Ω_N = Maximum Frequency of the arbitrary signals

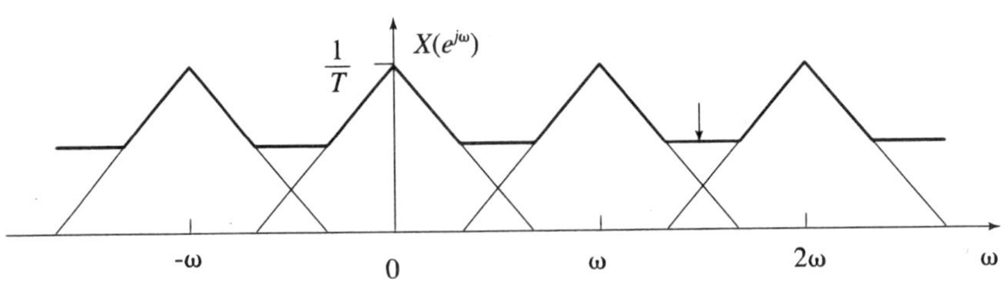

Figure 3-8 Aliasing effect

The following transformation relation between $x[n]$ and $X[k]$ is called DFT (Discrete Fourier Transform):

$$x[n] = \frac{1}{N}\sum_{k=0}^{N-1} X[k]\exp(jk\Omega_o n), \quad \left(\Omega_o = \frac{2\pi}{N}\right)$$

$$X[k] = \sum_{n=0}^{N-1} x[n]\exp(-jk\Omega_o n).$$

For DFT, a finite number (N) of samples are used both for time and frequency, so it can be processed with a digital system. In addition, the two forms are similar to each other. The FFT (Fast Fourier Transform) algorithm makes it easy to process digital signals.

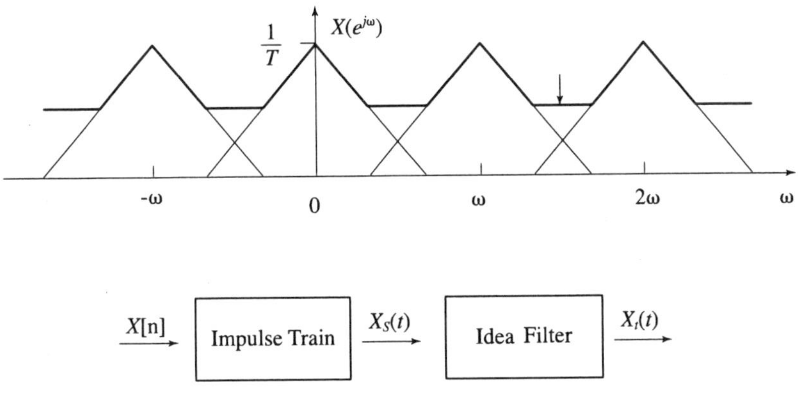

그림 3-9 Reconstruction of signal

In order to revert the discrete time signal acquired via sampling to the continuous time signal, it is required to transform the value to an impulse train. It is because this requires an impulse train on the contrary to acquiring a discrete time signal from a continuous time signal. The signal converted into an impulse train is reverted into the original continuous time signal via an appropriate low pass filter.

$$x_c(t)=\sum_{n=-\infty}^{\infty} x[n]\delta(t-nT)$$

The discrete time signal $x[n]$ becomes the impulse train through the above formula.

The original continuous time signal is made through an ideal low pass filter. Let's find out how to get the ideal low pass filter. The gain of the low pass filter (LPF) is T. Calculate the impulse response with the inverse Fourier transform (inverse CTFT) of the ideal LPF.

$$\begin{aligned} h_r(t) &= \frac{1}{2\pi}\int_{-\infty}^{\infty} H(j\Omega)e^{j\Omega t}d\Omega \\ &= \frac{T}{2\pi}\int_{-\pi/T}^{\pi/T} e^{j\Omega t}d\Omega \\ &= \frac{T}{2\pi}\left[\frac{e^{j\Omega t}}{jt}\right]^{\pi/T} \\ &= \frac{1}{2}\frac{e^{jt\pi/T}-e^{-jt\pi/T}}{jt\pi/T} \\ &= \frac{\sin\pi t/T}{(\pi/T)t} \end{aligned}$$

$x_r(t)$ can be recovered through convolution between $x_s(t)$ and the above impulse response. This is represented as follows:

$$\begin{aligned} x_r(t) &= \sum_{n=-\infty}^{\infty} x[n]\delta(t-nT)^{*}h_r(t) \\ &= \sum_{n=-\infty}^{\infty} x[n]h_r(t-nT) \\ &= \sum_{n=-\infty}^{\infty} x[n]\frac{\sin\sigma(t-nT)}{\sigma(t-nT)}, where\ \sigma=\frac{\pi}{T} \end{aligned}$$

Even the errorless perfect reconstruction signal has no ideal filter (because it is impossible to implement the SINC function in reality), and therefore, there occurs some data loss in reality.

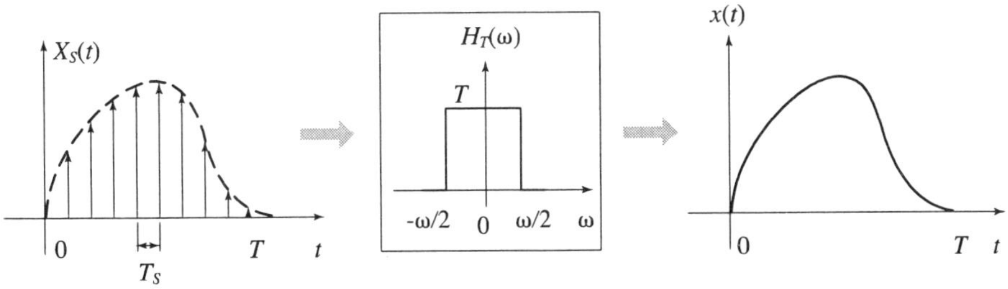

Figure 3-10 Reconstructed signal

Exercises

1. Calculate the energy of the following signal.

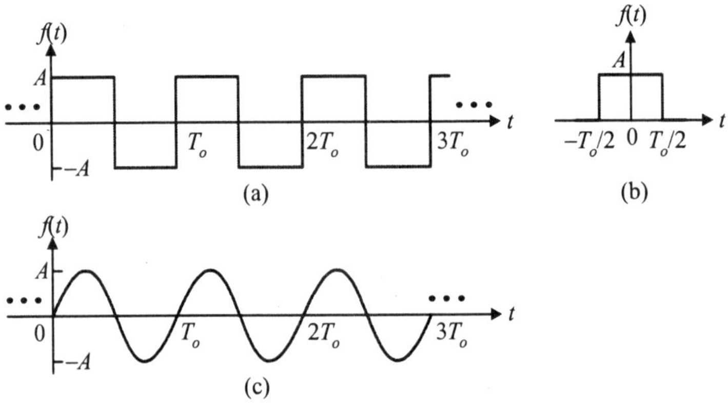

2. Calculate Fourier transform of the following impulse train:

$$\tilde{\delta}_T(t) \equiv \sum_{k=-\infty}^{\infty} \delta(t-kT)$$

3. Derive a time domain function of the following reconstruction filter:

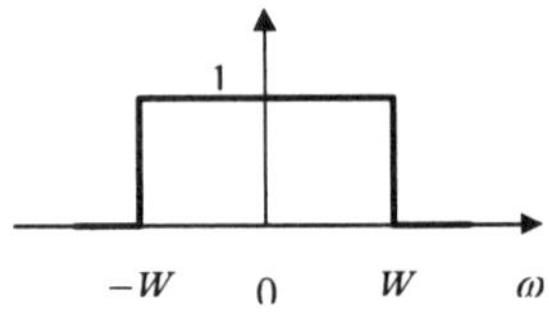

Answers

1. (a) : $E=\int_{-\infty}^{\infty} A^2 dt=\infty$

(b) : $E=\int_{-T_o/2}^{T_o/2} A^2 dt = A^2 T_o$

(a) : $E=\int_{-\infty}^{\infty} A^2 \sin^2\left(\frac{2\pi}{T_o}t\right) dt=\infty$

2. Calculate the exponential Fourier coefficient first.

$$c_n=\frac{1}{T}\int_{t_o}^{t_o+T} \tilde{\delta}_T(t)\exp(-jn\omega_o t)dt$$

$=\frac{1}{T}\int_{t_o}^{t_o+T} \delta(t)\exp(-jn\omega_o t)dt$ (Because there is only one impulse function in a period)

$$=\frac{1}{T}\int_{t_o}^{t_o+T} \delta(t)dt=\frac{1}{T}.$$

$$F[\tilde{\delta}_T(t)]=F\left[\frac{1}{T}\sum_{n=-\infty}^{\infty}\exp(jn\omega_o t)\right]$$

$$=\frac{2\pi}{T}\sum_{n=-\infty}^{\infty}\delta(\omega-n\omega_o)$$

$$=\omega_o\sum_{n=-\infty}^{\infty}\delta(\omega-n\omega_o).$$

Chapter 4

Electric Field/Magnetic Field

In parallel plate electrodes shown in the following figure, where the area of the electrode is $S\,[\mathrm{m}^2]$, the electric charge at the electrode is $Q\,[\mathrm{C}]$, the electric flux density between electrodes is $D\,[\mathrm{C/m}^2]$, and the electric flux density on the space between electrodes is $D = Q/S\,[\mathrm{C/m}^2]$. If the space between electrodes is filled with dielectric substance with the permittivity of ϵ, the electric current I [A] and the electric flux density $J_D\,[\mathrm{A/m}^2]$ are calculated as follows:

$$I = \frac{dQ}{dt} = \frac{d}{dt}(\mathrm{D}\cdot\mathrm{S}) = \frac{d}{dt}(\epsilon\mathrm{E}\cdot\mathrm{S}) = S\cdot\frac{dD}{dt}$$

$$J_D = \frac{I}{S} = \frac{\left(S\cdot\frac{dD}{dt}\right)}{S} = \frac{dD}{dt} = \epsilon\frac{dE}{dt}\,[\mathrm{A/m}^2]$$

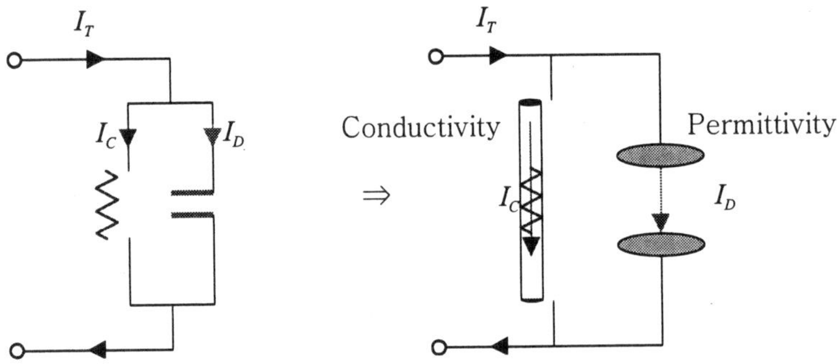

Figure 4-1 Displacement current between parallel plate electrodes

Maxwell assumed that current density exists in the dielectric substance between electrodes, and called this a displacement current. He thought that displacement current, like the current in the conductor (conduction current or characteristic current), creates a magnetic field. Therefore, the total current density J_T in a circuit is the sum of displacement current J_D and conduction current J_C.

$$J_T = J_C + J_D = \frac{1}{S}\frac{dQ}{dt} + \epsilon\frac{dE}{dt}\,[\mathrm{A/m^2}]$$

Where the polarization by electric field E[V/m] is P[C/m2] when the pole plate is filled with dielectric substance, the displacement current J_D in the dielectric substance is the sum of the displacement current generated by the change of electric field in a vacuum and the polarization current generated by movement of polarization charges.

$$J_D = \frac{dD}{dt} = \frac{d(\epsilon_o E + P)}{dt} = \epsilon_0\frac{dE}{dt} + \frac{dP}{dt}\,[\mathrm{A/m^2}]$$

1. Fundamental equation of electric field

Electromotive force e [V] induced by the current of the coil according to the Faraday's law:

$$e = -\frac{d\phi}{dt}\,[\mathrm{V}]$$

Electromotive force is equal to potential difference which is the difference of electric potential between two points so that change of potential energy in an electric field is Voltage.

$$e = V_{AB} = V_A - V_B$$
$$= \int_A^B E d\ell$$

Apply Stoke's theorem to this formula, and this is converted as follows:

$$\oint_\ell E d\ell = \int_S rot E ds = \int_S (\nabla \times E) ds$$
$$= -\frac{d\phi}{dt}\,[\mathrm{V}]$$

Divide both sides by $S = \int ds$.

$$\nabla \times E = -\frac{\partial \phi}{\partial t} \cdot \frac{1}{S} = -\frac{\partial(\phi/S)}{\partial t} = -\frac{\partial B}{\partial t}$$

$$\therefore \nabla \times E = -\frac{\partial B}{\partial t}$$

The left side is the rotation of the electric field, and the right side is the decreasing rate of the magnetic flux density over time. This means that if the magnetic flux density of a point in a space changes over time, there occurs a rotating electric field on the plane perpendicular to the magnetic flux density of the point. The direction of rotation of the electric field follows the right hand rule along the direction of the current. If there is no change in magnetic flux density, the right side is 0. In this case, the electromagnetic field becomes an irrotational field. In other words, there is no rotation of the electric field.

2. Fundamental equation of magnetic field

When calculating the Stoke's theorem by substituting the magnitude of magnetic field H[AT/m] calculated with the current I[A] given in the Ampere's circuital law, the following relational expression is established between the magnitude of magnetic field and the current density:

$$I = \oint_{\ell} H \cdot d\ell = \int_{S} rot\, H \cdot ds = \int_{S} \nabla \times H \cdot ds$$

$$= \int_{S} J \cdot ds = \int_{S} (J_C + J_D) \cdot ds = \int_{S} (J_C + \frac{dD}{dt}) \cdot ds$$

$$\therefore \nabla \times H = J_C + \frac{dD}{dt}$$

The left side is the rotation of the magnetic field, while the right side is the sum of the conduction current density and displacement current density. This means that if there is conduction current and displacement current at a point in a space, a magnetic vortex occurs centering on the point on the plane perpendicular to the resultant vector between the conduction current density and the displacement current density. The direction of rotation of the magnetic field follows the right handed rule along the direction of the resultant vector. In a medium with no conductivity, like dielectric substance, the magnetic field occurs with the displacement current only.

3. Divergence theorem (Gauss's law)

From Gauss's law that defines the total electric charges ($\sum Q$[C]) and electric flux density (D[C/m^2]) inside a closed surface in the electric field where the permittivity is ϵ [F/m], the relation between volume charge density (ρ_v [C/m^3]), electric flux density and magnitude of electric field (E[V/m]) is as follows:

$$\sum Q = \oint_S \epsilon E \cdot nds = \oint_S D \cdot nds$$

$$= \int_V div D ds = \int_V \nabla \cdot D ds$$

$$= \int_V \rho_v \, dv$$

$$\therefore \nabla \cdot D = \rho_v$$

This means that divergence of electric flux density equals charge density. In other words, in a space with charges, electric flux density that is equal to the volume charge densityρ_v is created ($\rho_v > 0$) or becomes extinct ($\rho_v < 0$) at a point. If there is no charge at this point ($\rho_v = 0$), no divergence exists at the point, and the continuous condition for no creation or extinction is set up. In this condition, there is no change of electric flux density, and no increase or decrease of electric flux density.

In the same way, it is possible to explain the relation between total magnetic charges ($\sum m$ [Wb]), magnetic flux density (B[Wb/m^2]) and magnitude of magnetic field (H[AT/m]) inside the closed surface in the magnetic field of the space with the magnetic permeability of μ [H/m]. The total magnetic charge is 0 because it is impossible to a split differential magnet (magnetic domain) within a magnetic field.

$$\sum m = \oint_S \mu H \cdot nds = \oint_S B \cdot nds$$

$$= \int_V div B \, ds = \int_V \nabla \cdot B \, ds$$

$$= 0$$

$$\therefore \nabla \cdot B = 0$$

This means that the divergence of magnetic flux density is always 0. Unlike the electric field, the divergence of a magnetic field is always 0 because there is no magnetic charge in the magnetic field which is equivalent to electric charge. If the divergence is 0, no magnetic flux density is created or becomes extinct at any point in a space. Therefore, magnetic flux or magnetic line of force is always continuous and becomes a closed curve. The space with the divergence of 0 is called a sourceless field or a solenoidal field.

The above 4 formulas are Maxwell equations. The following table lists the Maxwell's equations in differentiation and integration:

Table 4-1 Maxwell's equations

	Differential form	Integral form	Related laws
General form	$\nabla \times H = J_c + \frac{\partial D}{\partial t}$	$\oint H \cdot dl = \int_s \left(J_c + \frac{\partial D}{\partial t}\right) \cdot ds$	Ampere′s law
	$\nabla \times E = -\frac{\partial B}{\partial t}$	$\oint E \cdot dl = \int_s \left(-\frac{\partial B}{\partial t}\right) \cdot ds$	Faraday′s law
	$\nabla \cdot D = \rho$	$\oint_s D \cdot ds = \int_v p \cdot dv$	Gauss′s law
	$\nabla \cdot B = 0$	$\oint_s B \cdot ds = 0$	No isolated magnetic charge
Free space (No electric charge or voltage)	$\nabla \times H = \frac{\partial D}{\partial t}$	$\oint H \cdot dl = \int_s \left(\frac{\partial D}{\partial t}\right) \cdot ds$	Ampere′s law
	$\nabla \times E = -\frac{\partial B}{\partial t}$	$\oint E \cdot dl = \int_s \left(-\frac{\partial B}{\partial t}\right) \cdot ds$	Faraday′s law
	$\nabla \cdot D = 0$	$\oint_s D \cdot ds = 0$	Gauss′s law
	$\nabla \cdot B = 0$	$\oint_s B \cdot ds = 0$	No isolated magnetic charge

Maxwell's equations explain the fact that time change of electric field makes rotation of magnetic field, and time change of magnetic field makes rotation of electric field. An oscillatory electric field makes rotation of an oscillatory magnetic field, and an oscillatory magnetic field makes rotation of an oscillatory electric field. An oscillatory electric field combined with an oscillatory magnetic field creates electromagnetic waves.

On the exterior of a conductor where the medium has no conductivity, no conduction current flows. Therefore, the following fundamental formula for electromagnetic waves is set up.

$$rot\ H = \nabla \times H = \frac{\partial D}{\partial t} = \epsilon \frac{\partial E}{\partial t},\quad Rot\ E = \nabla \times E = -\frac{\partial B}{\partial t} = -\mu \frac{\partial H}{\partial t}$$

① If $E(D)$ of a point changes over time, the point has a rotation component of H.

- H creates a small closed curve that links with the changing $E(D)$.
- As $E(D)$ changes by time, H also changes over time.

$$\nabla \times H = \frac{\partial D}{\partial t}$$

② If $H(B)$ of a point changes over time, the point has a rotation component of E.

- E creates a small closed curve that links with the changing $H(B)$.
- As $H(B)$ changes over time, E also changes over time.

$$\nabla \times E = -\frac{\partial B}{\partial t}$$

③ As the two states are repeated, the current flows at the light velocity.

In the free space with no conduction current ($J_c = 0$), the electric field is assumed to oscillate in the direction of x only in parallel to the $x-z$ plane.

- The magnetic field oscillates in the direction of y only in parallel to the $y-z$ plane.
- The electric field and the magnetic field form a uniform plane wave.

In the non-conductive medium (ϵ [F/m], μ [H/m] conductivity = 0, True electric charge density = 0, current density= 0), the following formula is set up as described in Maxwell's equations:

$$\nabla \times H = \epsilon \frac{\partial E}{\partial t}, \quad \nabla \times E = -\mu \frac{\partial H}{\partial t}, \quad \nabla \cdot E = 0, \quad \nabla \cdot B = 0$$

This formula is electric field E[V/m] and magnetic field H[A/m] only, and has the following wave characteristics. Applying vector rotation operation for rotation component of the electric field, and substituting rotation of electric field and divergence of electric field will make the following formula.

$$\nabla \times \nabla \times E = \nabla(\nabla \cdot E) - \nabla^2 E = -\frac{\partial(\nabla \times B)}{\partial t}$$

$$-\nabla^2 E = \mu \frac{\partial(\nabla \cdot H)}{\partial t} = -\mu \frac{\partial^2 D}{\partial t^2}, \quad \frac{\partial^2 E_x}{\partial z^2} = \mu\epsilon \frac{\partial^2 E_x}{\partial t^2}$$

$$\because \nabla \cdot D = 0, \ D = \epsilon E, \ B = \mu H, \ \nabla \times H = \frac{\partial D}{\partial t}$$

Applying vector rotation operation for rotation component of the magnetic field, and substituting rotation of magnetic field and divergence of magnetic field will make the following formula:

$$\nabla \times \nabla \times H = \nabla(\nabla \cdot H) - \nabla^2 H = -\frac{\partial(\nabla \times D)}{\partial t}$$

$$-\nabla^2 H = \epsilon \frac{\partial(\nabla \times H)}{\partial t} = -\epsilon \frac{\partial^2 B}{\partial t^2}, \quad \frac{\partial^2 H_y}{\partial z^2} = \mu\epsilon \frac{\partial^2 H_y}{\partial t^2}$$

$$\because \nabla \cdot B = 0, \ B = \mu H, \ D = \epsilon E, \ \nabla \times E = -\frac{\partial B}{\partial t}$$

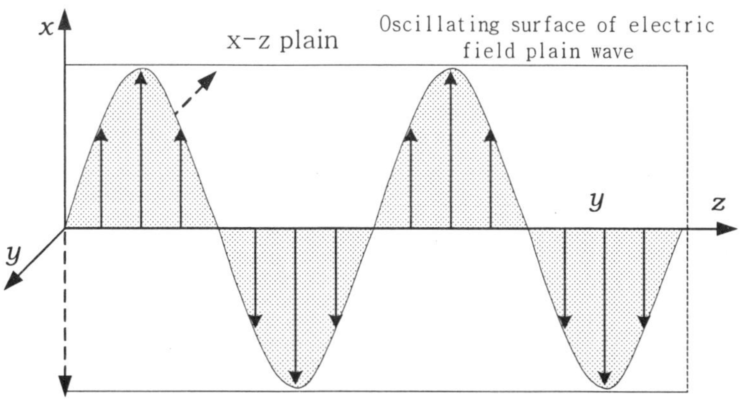

Figure 4-2 Plain wave of electric field

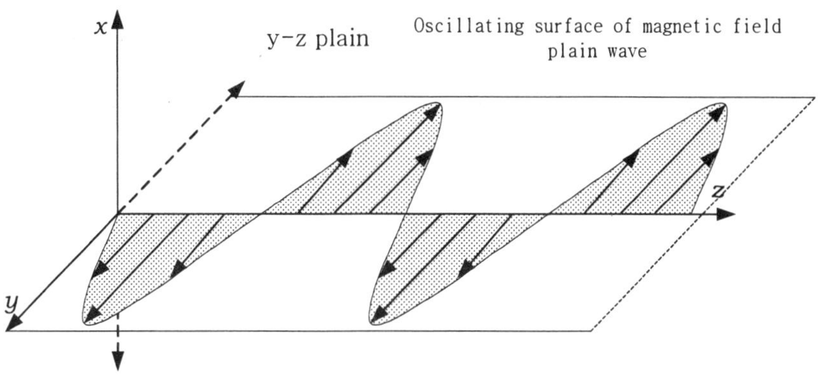

Figure 4-3 Plain wave of magnetic field

This is called a wave equation. Because there is no z-direction in a uniform medium, the wave equation can be rewritten as follows:

$$\epsilon \frac{\partial Ex}{\partial t} = -\frac{\partial Hy}{\partial Z}$$

$$-\mu \frac{\partial Hy}{\partial t} = \frac{\partial Ex}{\partial Z}$$

$$\epsilon \frac{\partial Ey}{\partial t} = \frac{\partial Hx}{\partial Z}$$

$$\mu\frac{\partial Hx}{\partial t}=-\frac{\partial Ey}{\partial Z}$$

The general solution of this formula is represented as follows, using functions f_x, f_y, g_x and g_y :

$$E_x=f_x(z-vt)+g_x(z+vt)$$

$$H_y=f_y(z-vt)+g_y(z+vt)$$

Where, $v=1/\sqrt{\epsilon\mu}=3\times10^8/\sqrt{\epsilon_s\mu_s}$ [m/sec], and $\epsilon_s=1$ and $\mu_s=1$ in a vacuum, and the speed is equivalent to speed of light. This is called a propagation velocity. $f_x(z-vt)$ and $f_y(z-vt)$ are the traveling wave at the velocity of v to the direction of $+z$, while $g_x(z+vt)$ and $g_y(z+vt)$ are the reflecting wave at the velocity of v to the direction of -z. In other words, a TEM (transverse electro magnetic) wave consisting of $+z$ forward wave and $-z$ reverse wave is made.

For the $+z$ wave only,

$$E_x=f_x(z-vt)$$

$$E_y=f_y(z-vt)$$

$$H_x=-\sqrt{\frac{\epsilon}{\mu}}f_y(z-vt)$$

$$H_y=-\sqrt{\frac{\epsilon}{\mu}}f_x(z-vt)$$

Any moment can be represented on xy plane as follows:

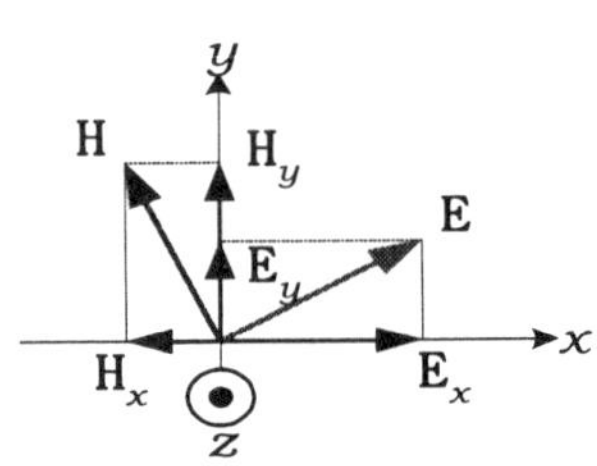

Figure 4-4 Phasor diagram of +z wave

$$\therefore Z_0 = \frac{E}{H} = \frac{\sqrt{E_x^2 + E_y^2}}{\sqrt{H_x^2 + H_y^2}} = \sqrt{\frac{\mu}{\epsilon}} = 120\pi\sqrt{\frac{\mu_s}{\epsilon_s}} = 377\sqrt{\frac{\mu_s}{\epsilon_s}}$$

This is called an intrinsic impedance or wave impedance.

The wave where the electric field and the magnetic field have no z component, and the planes on which electric fields and magnetic fields are perpendicular to z-axis is called a plane wave. The wave proceeding along these electric fields and magnetic fields is called an electro-magnetic wave or electric wave.

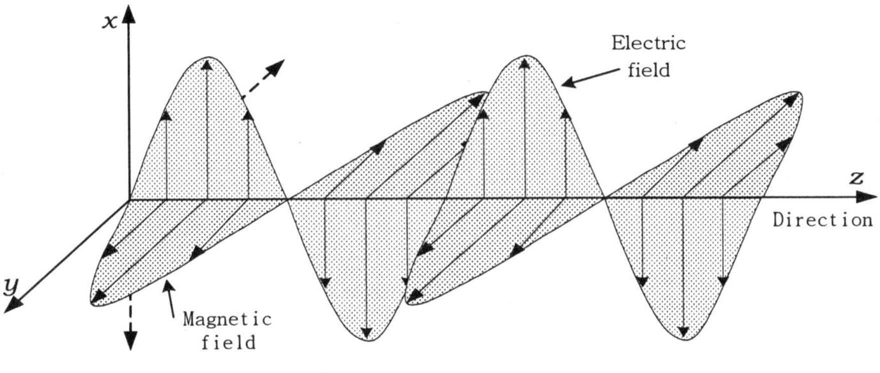

Figure 4-5 Progress of electromagnetic wave

Where the electric field and the magnetic field changes in the form of sine wave to the directions of x and y, and the electromagnetic wave to the direction of z has wavelength of $\lambda[m]$, frequency (f) of $\frac{C}{\lambda}$[Hz], and angular frequency (w) of $2\pi f$[rad],

$$E_x = f_z(z - vt)$$

$$= E_m \sin(wt - \frac{2\pi}{\lambda} \cdot z)$$

$$= E_m \sin w(t - \frac{z}{v})$$

Where, $\frac{2\pi}{\lambda}$ is called a phase factor β.

④ As the plane wave with the angular frequency w oscillates in the form of sine wave:

$$\begin{cases} E_x^+ = E_m \sin(wt - \beta z) \text{ : Incident wave} \\ E_x^- = E_m \sin(wt + \beta z) \text{ : Reflected wave} \end{cases} \quad \begin{cases} H_y^+ = H_m \sin(wt - \beta z) \text{ : Incident wave} \\ H_y^- = H_m \sin(wt + \beta z) \text{ : Reflected wave} \end{cases}$$

$\beta = w\sqrt{\mu\epsilon}\,[rad/m]$; Phase constant, $H_m = \sqrt{\frac{\epsilon}{\mu}}E_m \quad \therefore H_y = \sqrt{\frac{\epsilon}{\mu}}E_x$

⑤ Velocity of electromagnetic wave

$$v = \frac{1}{\sqrt{\mu\epsilon}} = \frac{1}{\sqrt{\mu_o\epsilon_o}} \cdot \frac{1}{\sqrt{\mu_s\epsilon_s}} = \frac{C_o}{\sqrt{\mu_s\epsilon_s}} = \frac{3\times10^8}{\sqrt{\mu_s\epsilon_s}}[\mathrm{m/sec}]$$

⑥ Intrinsic wave impedance

$$n = \frac{E}{H} = \sqrt{\frac{\mu}{\epsilon}} = \sqrt{\frac{\mu_o}{\epsilon_o}} \cdot \sqrt{\frac{\mu_s}{\epsilon_s}} = 120\pi\sqrt{\frac{\mu_s}{\epsilon_s}} = 377\sqrt{\frac{\mu_s}{\epsilon_s}} \quad [\Omega]$$

Characteristic impedance in vacuum $Z_o = \sqrt{\frac{\mu_o}{\epsilon_o}} = 120\pi = 377[\Omega]$

⑦ Wavelength

$$\lambda = \frac{v}{f} = \frac{1}{f\sqrt{\mu \cdot \epsilon}}[\mathrm{m}]$$

In electric field theory, the electric energy is $w_e = \frac{1}{2}ED[\mathrm{J/m^3}]$, and in magnetic field theory, the magnetic energy is $w_m = \frac{1}{2}BH[\mathrm{J/m^3}]$. If an electric field and a magnetic field coexist, the energy density of the electromagnetic field is calculated as follows,

$$w_o = w_e + w_m = \frac{1}{2}\epsilon E^2 + \frac{1}{2}\mu H^2[\mathrm{J/m^3}]$$

Because $H = \sqrt{\frac{\epsilon}{\mu}}E$, and $E = \sqrt{\frac{\mu}{\epsilon}} \cdot H$,

$$w_0 = \sqrt{\epsilon\mu}\, EH[\mathrm{J/m^3}]$$

This is the energy density of the average electromagnetic wave. A plane electromagnetic wave propagates at the speed of $v = 1/\sqrt{\epsilon\mu}\,[\mathrm{m/sec}]$ perpendicular to electric wave and magnetic wave. The energy flow P that passes a unit area which is perpendicular to the propagation direction in a unit of time is as follows:

$$P = w_0 \cdot v = \sqrt{\epsilon\mu} \cdot EH/\sqrt{\epsilon\mu} = EH[\mathrm{W/m^2}]$$

Because the electric field E and the magnetic field H are perpendicular to each other, the outer (cross) product is $P = E \times H$, and the direction of vector P corresponds with the direction of the electromagnetic wave. The electromagnetic energy passing the surface S per second is

$$P = \int P \cdot nds\,[\mathrm{W}]$$

This relation is called the Poynting theorem, and vector P is called a Poynting vector, or a radiation vector.

Part II

Chapter 5

Differential Equation

1. Differentiation

1) Differential coefficient

In a function $y=f(x)$, if variable x changes to $x_1+\Delta x$, the average change rate of this function is as follows:

$$\frac{\Delta y}{\Delta x}=\frac{f(x_1+\Delta x)-f(x_1)}{\Delta x} \tag{5-1}$$

If $\Delta x \to 0$, it is expressed as follows:

$$\lim_{\Delta x\to 0}\frac{\Delta y}{\Delta x}=\lim_{\Delta x\to 0}\frac{f(x_1+\Delta x)-f(x_1)}{\Delta x} \tag{5-2}$$

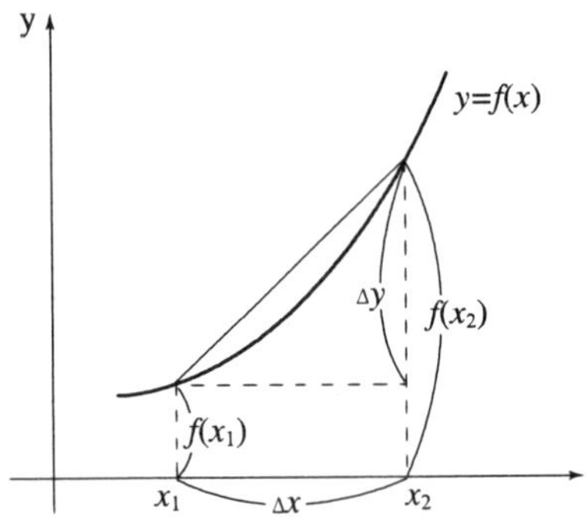

Figure 5-1 Differentiation of a function

If there is a limit value, $f(x)$ is differentiable in $x=x_1$. The limit value is called a differential coefficient for $x=x_1$ of the function $y=f(x)$.

$$f'(x_1)=\lim_{\Delta x\to 0}\frac{\Delta y}{\Delta x}=\lim_{\Delta x\to 0}\frac{f(x_1+\Delta x)-f(x_1)}{\Delta x} \tag{5-3}$$

Exercise 5-1

Evaluate $f'(3)$ in the function $f(x)=2x+1$.

Solution

$$f'(3)=\lim_{\Delta x\to 0}\frac{f(3+\Delta x)-f(3)}{\Delta x}=\lim_{\Delta x\to 0}\frac{2(3+\Delta x)+1-(2\cdot 3+1)}{\Delta x}$$

$$=\lim_{\Delta x\to 0}\frac{2\Delta x}{\Delta x}=\lim_{\Delta x\to 0}(2)=2$$

2) Derivatives

In $x=a$ of the function $y=f(x)=x^2$, the differential coefficient $f'(a)$ is as the following:

$$f'(a) = \lim_{\Delta x \to 0} \frac{(a+\Delta x)^2 - a^2}{\Delta x} = \lim_{\Delta x \to 0} (2a + \Delta x) = 2a \qquad (5\text{-}4)$$

Substitute $a = 1,\ 2,\ 3 \cdots$ for $f'(a) = 2a$.

$$f'(1) = 2,\ f'(2) = 4,\ f'(3) = 6, \cdots$$

The differential coefficient $f'(a_1)$ in the function $y = f(x)$ when $x = a$ is a function of a if a is a variable. Therefore, if x is substituted for a in $f'(a_1)$, $f'(x)$ is a function of x; this $f'(x)$ is called a derivative of $f(x)$.

$$f'(x) = \lim_{\Delta x \to 0} \frac{f(x+\Delta x) - f(x)}{\Delta x}$$

Derivatives are represented with the following symbols:

$$y',\ f'(x), \frac{dy}{dx}, \frac{d}{dx} f(x)$$

Evaluating the derivative $f'(x)$ in the function $y = f(x)$ is differentiating $f(x)$ for x.

Exercise 5-2

Differentiate the function $y = 2x^2 + 3$.

$$y' = \lim_{\Delta x \to 0} \frac{f(x+\Delta x) - f(x)}{\Delta x} = \lim_{\Delta x \to 0} \frac{2(x+\Delta x)^2 + 3 - (2x^2 + 3)}{\Delta x}$$

$$= \lim_{\Delta x \to 0} \frac{4x \cdot \Delta x + 2(\Delta x)^2}{\Delta x} = \lim_{\Delta x \to 0} (4x + 2\Delta x) = 4x$$

Solution

The following properties are required to evaluate the derivative:

i) $\{f(x) \pm g(x)\}' = f'(x) \pm g'(x)$

ii) $\{c \cdot f(x)\}' = c \cdot f'(x)$ (c is a constant)

iii) $\{f(x) \cdot g(x)\}' = f'(x)g(x) + f(x)g'(x)$

iv) If $f(x) = c$, $f'(x) = 0$

v) If $f(x)=x^n$, $f'(x)=nx^{n-1}$(n is a natural number)

vi) $\left\{\frac{f(x)}{g(x)}\right\}'=\frac{f'(x)g(x)-f(x)g'(x)}{\{g(x)\}^2}$

Exercise 5-3

Differentiate the following functions:

(1) $y=2x^3-4x^2+5$ (2) $y=x^2(x-2)$

Solution

(1) $y'=(2x^3-4x^2+5)'=(2x^3)'-(4x^2)'+(5)'$

$=2\cdot(x^3)'-4(x^2)'+0=2\cdot 3x^{3-1}-4\cdot 2x^{2-1}$

$=6x^2-8x$

(2) $y'=\{x^2(x-2)\}'=(x^2)'(x-2)+x^2(x-2)'$

$=2x(x-2)+x^2$

$=3x^2-4x$

Exercise 5-4

On the curve $f(x)=(x^2-1)(x+2)$, evaluate the derivatives at the point of $x=$ 1.

Solution

$f'(x)=\{(x^2-1)(x+2)\}'=(x^2-1)'(x+2)+(x^2-1)(x+2)'$

$=2x(x+2)+(x^2-1)=3x^2+4x-1$

Therefore, the gradient at the point of $x=1$ is as the following:

$f(1)=3\cdot 1^2+4\cdot 1-1=6$

2. Basic concept and family of curves

An equation with derivatives is called a differential equation. The followings are examples:

$$\frac{dy}{dx} = \cos x \tag{5-5}$$

$$\frac{d^2y}{dx^2} + 3\frac{dy}{dx} + 2y = 0 \tag{5-6}$$

$$\left(\frac{d^3y}{dx^3}\right)^2 + 2\left(\frac{d^2y}{dx^2}\right)^4 + \frac{dy}{dx} = (x^2+2)y^2 \tag{5-7}$$

$$\frac{\partial^4 y}{\partial x^4} = \frac{\partial^2 y}{\partial x \partial t} + xt\sin(yt) \tag{5-8}$$

Differential equations are classified into ordinary differential equations and partial differential equations. An equation with a single independent variable, like (5-5), (5-6) and (5-7), is called an ordinary differential equation, while the one with two independent variables, like (5-8), is called a partial differential equation.

The highest value variable of derivatives in a differential equation is called the order of the equation. Equations (5-5), (5-6), (5-7) and (5-8) are 1st, 2nd, 3rd and 4th-order differential equation, respectively.

In a differential equation, the degree of derivative of the highest order for an unknown function y and its derivative is called the degree of differential equation. Especially, the 1st-order equation for an unknown function y and its derivative is called a linear differential equation, and 2nd or higher order equation is called a nonlinear differential equation. In this section, derivatives $dy/dx, d^2y/dx^2, \cdots, d^ny/dx^n$ will be marked as y', y'', $\cdots y,^{(n)}$.

The function

$$y = g(x) \tag{5-9}$$

is a solution in the section $a < x < b$ for the following differential equation:

$$F(x,\ y,\ y') = 0 \tag{5-10}$$

For example, the function

$$y = g(x) = e^{2x}$$

is the solution of the following 1st-order differential equation:

$$y' - 2y = 0$$

It is because, 0=0 when the function is differentiated, and

$$g(x) = 2e^{2x} = y'$$

where $g(x) = y$ is substituted for the differential equation.

Now, let's find out the relation between the families of curves, which are widely used in natural science and engineering and differential equations.

$y = \sin x + c$ is the solution for Equation (5-4). As shown in Figure (5-2), a curve group is formed with a number of curves according to a constant c.

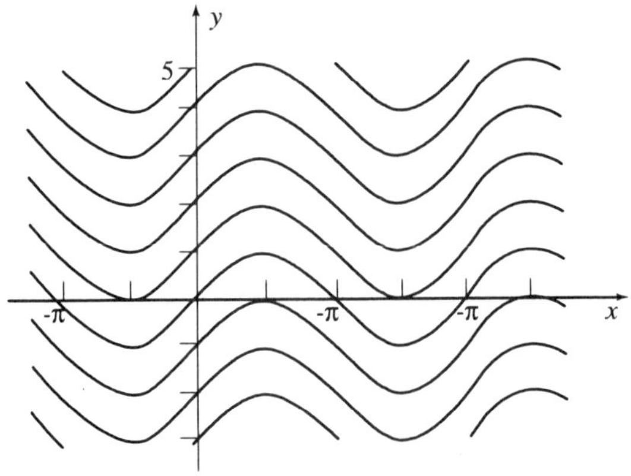

Figure 5-2 Families of curves

In general, the following equation with a constant c indicates a family of curves:

$$F(x,\ y,\ c)=0 \tag{5-11}$$

In order to find the common feature of all the curves of this family, take equation (5-11), partial-differentiate both sides of (5-11) with x, and get Equation (5-10) from the following equation:

$$\frac{\partial}{\partial x}F(x,\ y,\ c)+\frac{\partial}{\partial y}F(x,\ y,\ c)\cdot y'=0 \tag{5-12}$$

It is the 1st-order differential equation showing the properties of the family of curves equation (5-12).

Exercise 5-5

Evaluate the differential equation derived from the family of curves $y=ce^x$ in Figure (5-3).

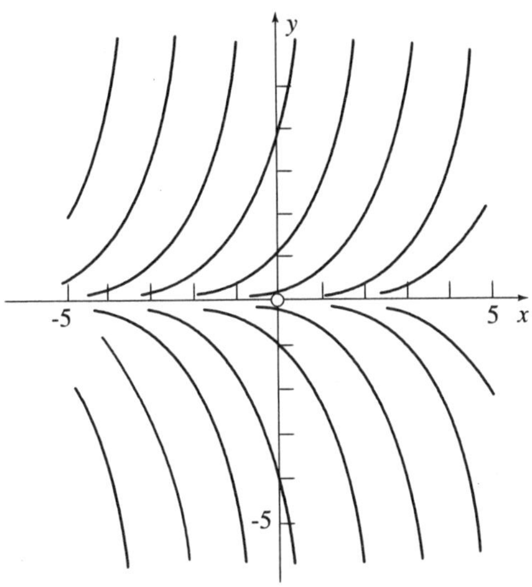

Figure 5-3 Family of curves

Solution

Differentiate the given equation $y=ce^x$, and we will get the following:

$y'=ce^x$

Substituting $c=\frac{y}{e^x}$, and we will get the following 1st-order differential equation:

$y'-y=0$

As we have seen in the above example, a 1st-order differential equation has the solution of a constant c. This solution is called a general solution. If this general solution is $F(x, y, c)=0$, substituting $c_1, c_2 \cdots$ for a constant c will always yield a specific solution $F(x, y, c_1)=0$, $F(x, y, c_2)=0, \cdots$. This solution is called a particular solution.

In some cases, there can be a solution other than the given differential which cannot be acquired by substituting a particular value for a constant c in the general solution. This solution is called a singular solution.

Exercise 5.6

Evaluate the result of following equation.

$$y'^2-xy'+y=0$$

Solution

The general solution is as the following:

$y=cx-c^2$

This indicates a family of curves of the straight line in Figure (5-4). Other solution is as the following:

$y=\frac{x^2}{4}$

Because this solution cannot be acquired by substituting any value for the constant c in the general solution $y=cx-c^2$, it is a singular solution.

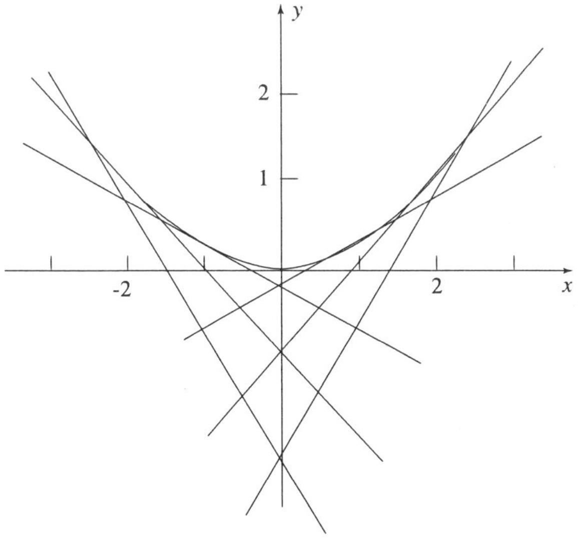

Figure 5-4 Family of curves of a straight line

As is apparent, each particular solution indicates a tangent line equation of an arc with the singular solution $y=\frac{x^2}{4}$.

Amonh the many applications of engineering mathematics, we are interested in a particular solution $y(x)$ which satisfies the initial condition $y(x_0)=y_0$, rather than the general solution of the given differential equation. The differential equation with an intial condition is called an initial value problem.

Exercise 5-7

The general solution of the differential equation $2xy'-y=0$ is $y^2=cx$. Find the particular solution if the initial condition is $y(1)=4$.

Solution

Substitute the initial condition for the general solution, and we will get $c=16$. Substitute this for the general solution, and we will get the particular solution

$y^2 = 16x$.

---

3. Separable equation

Many 1st-order differential equations can be transformed into the following:

$$g(y)y' = f(x) \tag{5-13}$$

Because $y' = \frac{dy}{dx}$, equation (5-13) can be transformed into the following:

$$g(y)dy = f(x)dx \tag{5-14}$$

This equation is called a separable equation. y is the only variable in the left side of Equation (5-14), and x in the right side. Therefore, integrating both sides will result in the following equation:

$$\int g(y)dy = \int f(x)dx + c \tag{5-15}$$

If f and g are continuous functions, there is an integration of equation (5-15). Calculate this integration to evaluate the general solution of equation (5-13).

Exercise 5-8

Solve the following differential equations:

(a) $9yy' + 4x = 0$ (b) $(1+x)y' = 1+y$

Solution

(a) Separate variables as the following:

$9ydy = -4xdx$

Integrate both sides, and we will get the following general solution:

$$\frac{9}{2}y^2 = -2x^2 + c' \text{ or } \frac{x^2}{9} + \frac{y^2}{4} = c\left(c = \frac{c'}{18}\right)$$

This solution indicates a family of ovals.

(b) Separate variables and integrate both sides, and we will get the following general solution:

$$\ln(1+y) = \ln(1+x) + c'$$

This can be transformed to

$$1+y = e^{c}(1+x) \text{ or } y = cx + (c-1) \; (c = e^{c}).$$

Exercise 5-9

Solve the following initial value problem:

$$(x^2+1)y' + y^2 + 1 = 0, \quad y(0) = 1$$

Solution

Separate variables.

$$\frac{dy}{1+y^2} = -\frac{dx}{1+x^2}$$

Integrate both sides.

$$\tan^{-1}y = -\tan^{-1}x + c \text{ or} \tan^{-1}y + \tan^{-1}x = c$$

$$\frac{y+x}{1-xy} = \tan c$$

Substitute the initial condition, and then, $\tan c = 1$. Therefore,

$$y = \frac{1-x}{1+x}.$$

4. Homogeneous equation

A 1st-order differential equation, which is not a separable equation, can become a separable equation through the transformation of a variable. In a differential equation $y' = g(x, y)$, the following 1st-order differential equation in which $g(x, y)$ is a function of y/x only is called a homogeneous equation:

$$y' = g\left(\frac{y}{x}\right) \tag{5-16}$$

The homogeneous differential equation (5-16) can become a separable equation through the following transformation of variables:

$$\frac{y}{x} = u \tag{5-17}$$

It becomes the following with differentiation of $y = ux$:

$$y' = u + u'x \tag{5-18}$$

If (5-17) and (5-18) are substituted for (5-16), it becomes the following equation:

$$u + u'x = g(u)$$

Separating variables to u and x will make the following:

$$\frac{du}{g(u) - u} = \frac{dx}{x}$$

Integrating this equation and substituting y/x for u will make the general solution of (5-16).

Exercise 5-10

Solve $2xyy' - y^2 + x^2 = 0$.

Solution

Divide both sides by x^2, and the following homogeneous equation is made:

$$2\frac{y}{x}y' - \left(\frac{y}{x}\right)^2 + 1 = 0$$

Substituting $u = y/x$ and $u + u'x = y'$ will make the equation following: $2u(u + u'x) - u^2 + 1 = 0$ or $2xuu' + u^2 + 1 = 0$

Separate variables to make the following:

$$\frac{2udu}{1+u^2}=-\frac{dx}{x}$$

Integrate it and we will get the following general solution:

$$\ln(1+u^2)=-\ln|x|+c' \text{ or } 1+u^2=\frac{c}{x}$$

Substituting y/x for u, and the following equation is made:

$$x^2+y^2=cx \text{ or} \left(x-\frac{c}{2}\right)^2+y^2=\frac{c^2}{4}$$

In certain types of differential equation, it is often required to perform transformation of variables according to the circumstance.

Exercise 5-11

Solve $y'=g\left(\dfrac{ax+by+c}{a'x+b'y+c'}\right)$.

where, a, b, c, a', b' and c' are constants.

Solution

(a) $\dfrac{a}{a'}\neq\dfrac{b}{b'}$

In this case, roots ξ and η of the following simultaneous equations are determined:

$$\begin{cases} ax+by+c=0 \\ a'x+b'y+c'=0 \end{cases}$$

Perform transformation of variables as follows:

$$x=u+\xi,\ y=v+\eta$$

And the following homogeneous equation is made:

$$\frac{dv}{du}=g\left(\frac{au+bv}{a'u+b'v}\right)$$

(b) $\dfrac{a}{a'}=\dfrac{b}{b'}$

In this case, assume this rate as $1/k$, and the following is made:

$$y'=g\left[\frac{ax+by+c}{k(ax+by)+c'}\right]$$

If $ax+by=u$,

$$y'=\frac{1}{b}\left(\frac{du}{dx}-a\right).$$

Therefore, the equation is a separable equation:

$$\frac{du}{dx}=a+bg\left(\frac{u+c}{ku+c'}\right)$$

Exercise 5-12

Solve the following 1st-order differential equation:

(a) $(2x-4y+5)y'+x-2y+3=0$

(b) $(-2x-2y+1)y'+6x-2y-3=0$

Solution

(a) Because x,y coefficients are proportional, it is the case of (b) in Exercise 11. If $x-2y=u$, then, $y'=\frac{1}{2}(1-u')$. Substituting it in the equation will result in the following:

$$(2u+5)u'=4u+11$$

Separate variables and integrate both sides, and we will get the following general solution:

$$\left(1-\frac{1}{4u+11}\right)du=2dx,\ u-\frac{1}{4}ln|4u+11|=2x+c^*$$

Substitute $u=x-2y$, and then,

$$4x+8y+\ln|4x-8y+11|=c$$

(b) It is the case of Exercise 11(a). The solution of the following simultaneous equations is $\xi=\frac{1}{2}$, $\eta=0$:

$$\begin{cases}6x-2y-3=0\\-2x-2y+1=0\end{cases}$$

If $x=u+\frac{1}{2}$, then, $dx=du$. Substituting it in the equation will result in the following homogeneous equation:

$$(3u-y)dy-(u+y)dy=0, \text{ that is, } \frac{dy}{du}=\frac{3-\frac{y}{u}}{1+\frac{y}{u}}$$

Therefore, if $y=vu$, then it becomes the following:

$$v+u\frac{dv}{du}=\frac{3-v}{1+v}, \text{ that is, } \frac{du}{u}+\frac{(v+1)dv}{v^2+2v-3}=0$$

Integration will result in the general solution of the following:

$$u\sqrt{v^2+2v-3}=c^*$$

Substituting $u=x+\frac{1}{2}$, $v=\frac{y}{u}=\frac{y}{x+\frac{1}{2}}$ will make the following:

$$3x^2-2xy-y^2-3x+y+c=0 \left(c=c^{*2}+\frac{3}{4}\right)$$

5. Exact differential equation

A 1st-order differential equation may be written as follows:

$$\boldsymbol{M}(x,\ y)dx + \boldsymbol{N}(x,\ y)dy = 0 \tag{5-19}$$

If the left side of this equation is the following total differential of a function $u(x,\ y)$, (5-19) is called an exact differential equation:

$$du = \frac{\partial_u}{\partial x}dx + \frac{\partial u}{\partial y}dy \tag{5-20}$$

If (5-19) is an exact differential equation, it can be written as follows:

$$du = 0$$

When integrated, the general solution of the following is made:

$$u(x,\ y) = c \tag{5-21}$$

Compare between (5-19) and (5-20):

$$(a)\ \frac{\partial u}{\partial x} = \boldsymbol{M},\ (b)\ \frac{\partial u}{\partial y} = \boldsymbol{N} \tag{5-22}$$

If$\boldsymbol{M}$ and $\boldsymbol{N}$ are defined, and derivatives are in series in a given section, then the following equations are made from (5-22):

$$\frac{\partial M}{\partial y} = \frac{\partial^2 u}{\partial x \partial y},\ \frac{\partial N}{\partial x} = \frac{\partial^2 u}{\partial y \partial x}$$

Based on the assumption of continuity, we can say that the 2nd derivatives are the same.

$$\frac{\partial M}{\partial y} = \frac{\partial N}{\partial x} \tag{5-23}$$

(5-23) is a necessary and sufficient condition for differential equation (5-19) to be an exact differential equation. $u(x,\ y)$ can be evaluated as follows:

From (5-22(a)), the following equation is made:

$$u=\int Mdx+k(y) \tag{5-24}$$

Where y is considered a constant and $k(y)$ plays a role of constant of integration. To determine $k(y)$, derive $\partial u/\partial y$ from (5-24), and evaluate and integrate dk/dy by using (5-22(b)).

Exercise 5.13

Solve equation $xy'+y+4=0$·

Solution

This equation can be written as (5-19) in the following way:

$(y+4)dx+xdy=0$

Because the equation (5-23) $\frac{\partial M}{\partial y}=1=\frac{\partial N}{\partial x}$ is made, it is an exact differential equation. Integrate $N=\partial u/\partial y=x$ from (5-22(b)) to derive $u=xy+k(x)$.

In order to determine $k(x)$, differentiate u and apply (5-22(a)).

$\frac{\partial u}{\partial x}=y+\frac{dk}{dx}=M=y+4$

$dk/dx=4$, or $k=4x+c'$. Apply this to $u=xy+k$(=constant), and evaluate y. The general solution is

$y=\frac{c}{x}-4.$

Exercise 5.14

Solve equation $2x\sin 3ydx+3x^2\cos 3ydy=0$·

Solution

$\frac{\partial M}{\partial y}=6x\cos 3y=\frac{\partial N}{\partial x}$ from (5-23), and the equation is an exact differential equation. From (5-24), the following equation is made:

$u = \int 2x \sin 3y dx + k(y) = x^2 \sin 3y + k(y)$

$\partial u / \partial y = 3x^2 \cos 3y + dk/dy = N = 3x^2 \cos 3y$, and then, $k = c^* =$constant.

The general solution is u =constant, or $x^2 \sin 3y = c$.

6. Integrating factor

The following differential equation is not an exact differential equation:

$$P(x, y)dx + Q(x, y)dy = 0 \tag{5-25}$$

But the following equation derived by multiplying both sides with an appropriate function $F(x, y)(\neq 0)$ can be an exact differential equation:

$$F(x, y)P(x, y)dx + F(x, y)Q(x, y)dy = 0 \tag{5-26}$$

This function $F(x, y)$ is called an integrating factor.

Exercise 5.15

Solve $xdy - ydx = 0$.

Solution

Multiplying both sides by an integrating factor $F = 1/x^2$ will make the following:

$F(x)(xdy - ydx) = \frac{xdy - ydx}{x^2} = d\left(\frac{y}{x}\right) = 0,\ y = cx$

Multiplying both sides by $1/xy$ will make the following exact differential equation.

$\frac{1}{x}dx - \frac{1}{y}dy = 0,$ or $d\left(\ln\frac{x}{y}\right) = 0$

The general solution is $y=cx$, and $1/xy$ is an integrating factor.

As we have seen in Exercise 15, a differential equation has more than one integrating factor. There is no typical way to find integrating factors, but we need to rather experiment with a variety of formulas in order to find any integrating factors.

The following formulas may be used appropriately:

(1) $ydx+xdy=d(xy)$

(2) $ydx \pm xdy=\frac{1}{2}d(x^2 \pm y^2)$

(3) $\frac{xdy-ydx}{x^2}=d\left(\frac{y}{x}\right)$, $\frac{ydx-xdy}{y^2}=d\left(\frac{x}{y}\right)$

(4) $\frac{xdy-ydx}{x^2+y^2}=d\left(\tan^{-1}\frac{y}{x}\right)$, $\frac{ydx-xdy}{x^2+y^2}=d\left(\tan^{-1}\frac{x}{y}\right)$

(5) $\frac{xdy-ydx}{x^2-y^2}=\frac{1}{2}d\left(\ln\frac{x-y}{x+y}\right)$

(6) $\frac{xdy-ydx}{(x-y)^2}=\frac{1}{2}d\left(\frac{x+y}{x-y}\right)$, $\frac{ydx-xdy}{(x+y)^2}=\frac{1}{2}d\left(\frac{x-y}{x+y}\right)$

Exercise 5.16

Solve $xdy-ydx-2(x^2+y^2)dx=0$.

Solution

In the formula (4), we can find that $F=\frac{1}{x^2+y^2}$. Multiply both sides by this, and it becomes the following:

$\frac{xdy-ydx}{x^2-y^2}-2dx=0$, or $d\left(\tan^{-1}\frac{y}{x}-2x\right)=0$

The general solution is

$\tan^{-1}\frac{y}{x}-2x=c.$

It is very difficult to find an integrating factor $F(x, y)$. The following example can be used to find an integrating factor:

(a) If $\dfrac{\frac{\partial P}{\partial y}-\frac{\partial Q}{\partial x}}{Q}=f(x)$ is a function of x only, then, $F=e^{\int f(x)dx}$,

If $\dfrac{\frac{\partial P}{\partial y}-\frac{\partial Q}{\partial x}}{P}=-g(y)$ is a function of y only, $F=e^{\int g(x)dy}$ is an integrating factor of (5-25).

(b) If (5-25) is an homogeneous equation and $Px+Qy\neq 0$, $\dfrac{1}{Px+Qy}$ is an integrating factor.

(c) If (5-25) can be written as $yf(x)dx+xg(xy)dy=0$, $\dfrac{1}{Px+Qy}$ is an integrating factor.

Exercise 5.17

Not all the following differential equations are exact differential equations. Evaluate integrating factors and solve the equations.

(a) $(x^2+y^2+x)dx+xydy=0$

(b) $y^2dx+(x^2-xy-y^2)dy=0$

(c) $y(x^2y^2+2)dx+x(2-2x^2y^2)dy=0$

Solution

(a) Because $\dfrac{\partial P}{\partial y}=2y$, $\dfrac{\partial Q}{\partial x}=y$, it is not an exact differential equation.

$\dfrac{\frac{\partial P}{\partial y}-\frac{\partial Q}{\partial x}}{Q}=\dfrac{2y-y}{xy}=\dfrac{1}{x}=f(x)$, and then, the integrating factor is

$$F = e^{\int f(x)dx} = e^{\ln x} = x.$$

Multiply both sides by this integrating factor, and we will get the following:

$$(x^3 + xy^2 + x^2)dx + x^2ydy \text{ or} (x^3 + x^2)dx + d\left(\frac{1}{2}x^2y^2\right) = 0$$

The general solution is $3x^4 + 4x^3 + 6x^2y^2 = c$.

(b) The equation is homogeneous, and $\frac{1}{Px + Qy} = \frac{1}{y(x^2 - y^2)}$ is the integrating factor. Multiplying both sides of the equation by this integrating factor will make the following equation:

$$\frac{y}{x^2 - y^2}dx + \frac{x^2 - xy - y^2}{y(x^2 - y^2)}dy = 0$$

This is an exact differential equation (5-19). From (5-24), the following equation is derived:

$$u(x, y) = \int \frac{y}{x^2 - y^2}dx + k(y) = \frac{1}{2}\int\left(\frac{1}{x - y} - \frac{1}{x + y}\right)dx + k(y)$$

$$= \frac{1}{2}ln\frac{x - y}{x + y} + k(y) = \text{constant}$$

Differentiate this with y:

$$\frac{\partial u}{\partial y} = -\frac{x}{x^2 - y^2} + k(y) = N = \frac{y^2 - xy - y^2}{y(x^2 - y^2)} = \frac{1}{y} - \frac{x}{x^2 - y^2}$$

$k'(y) = \frac{1}{y}$, or $k(y) = \ln y$. Therefore, the general solution is

$$\frac{1}{2}ln\frac{x - y}{x + y} + \ln y = \ln c \text{ or } (x - y)y^2 = c(x + y).$$

(c) Because this equation is $yf_1(xy)dx + xf_2(xy)dy = 0$, the integrating factor is

$$\frac{1}{Px - Qy} = \frac{1}{3x^3y^3}.$$

Multiplying both sides of the equation by this integrating factor will make the following exact differential equation of (5-19):

$$\frac{x^2y^2 + 2}{3x^3y^2}dx + \frac{2 - 2x^2y^2}{3x^2y^3}dy = 0$$

From (5-24), the following equation is made:

$$u(x, y) = \int\left(\frac{x^2y^2 + 2}{3x^2y^2}\right)dx + k(y) = \int\left(\frac{1}{3x} + \frac{2}{3x^2y^2}\right)dx + k(g)$$

$$=\frac{1}{3}lnx-\frac{1}{3x^2y^2}+k(y)$$

Differentiate this with y:

$$\frac{\partial u}{\partial y}=\frac{2}{3x^2y^3}+k(y)=N=\frac{2-2x^2y^2}{3x^2y^3}$$

$$k'(y)=-\frac{2}{3y},$$

$$k(y)=-\frac{2}{3}lny.$$

Therefore, the general solution is

$$\frac{1}{3}lnx-\frac{1}{3x^2y^2}-\frac{2}{3}lny=\ln c \text{ or } x=cy^2e^{1/x^2y^2}$$

---

7. Linear 1st-order differential equation

For the following linear 1st-order differential equation:

$$\frac{dy}{dx}+f(x)y=0 \tag{5-27}$$

the general solution is

$$y(x)=ce^{-\int f(x)dx} \tag{5-28}$$

Change the constant c to an x function $c(x)$, and we can solve the following linear 1st-order differential equation:

$$\frac{dy}{dx}+f(x)y=r(x) \tag{5-29}$$

Substitute

$$y(x)=c(x)e^{-\int f(x)dx} \tag{5-30}$$

to the above (5-29).

$$c'(x)e^{-\int f(x)dx} = r(x) \quad (5\text{-}31)$$

$$c(x) = \int r(x)e^{\int f(x)dx}dx + c \qquad (5\text{-}32)$$

$$y(x) = e^{-\int f(x)dx}\left\{\int r(x)e^{\int f(x)dx}dx + c\right\} \qquad (5\text{-}33)$$

Because (5-33) has a constant c, it is the general solution of (5-29).

Exercise 5.18

Solve $y' + y = x^2$.

Solution

As in (5-33), the general solution is

$$y = e^{-\int dx}\left[\int x^2 e^{\int dx}dx + c\right]$$
$$= e^{-x}\left[\int x^2 e^x dx + c\right]$$
$$= e^{-x}[e^x(x^2 - 2x + 2) + c]$$
$$= x^2 - 2x + 2 + ce^{-x}$$

Exercise 5.19

Solve $(1+x^2)y' = xy + 1$.

Solution

The equation is transformed into the following:

$$y' - \frac{x}{1+x^2}y = \frac{1}{1+x^2}$$

Using (5-33), the following solution is made:

$$y = e^{\int \frac{x}{1+x^2}dx}\left[\int \frac{1}{1+x^2}e^{-\int \frac{x}{1+x^2}dx}dx + c\right]$$

$$= \sqrt{1+x^2}\left[\int \frac{dx}{(1+x^2)\sqrt{1+x^2}} + c\right]$$

$$= \sqrt{1+x^2}\left[\frac{x}{(1+x^2)} + c\right]$$

$$= x + c\sqrt{1+x^2}$$

The following Bernoulli differential equation is not a linear equation:

$$\frac{dy}{dx} + f(x)y = r(x)y^n \qquad (5\text{-}34)$$

If substituted by $z = y^{1-n}$, then $z' = (1-n)y^{-n}y'$, and the equation (5-34) becomes the following:

$$z' + (1-n)f(x) = (1-n)r \qquad (5\text{-}35)$$

which is the same as (5-29).

Exercise 5.20

Solve $y' + \frac{y}{x} = x^2y^3$.

Solution

This is a Bernoulli differential equation ($n=3$).

If $z = y^{1-3} = y^{-2}$, then

$z' = -2y^{-3} \cdot y'$.

Multiply both sides by $-2x^{-3}$ to make

$z' - \frac{2}{x}z = -2x^2$.

As in (5-33),

$$z = e^{\int \frac{2}{x}dx}\left[\int (-2x^2)e^{-\int \frac{2}{x}dx}dx + c\right]$$

$$= x^2\left[-2\int x^2\left(\frac{1}{x^2}\right)dx + c\right]$$

$$= -2x^3 + cx^2$$

Substitute $z = y^{-2}$, and then,

$$-2x^3y^2 + cx^2y^2 = 1.$$

Chapter 6

Complex Number

1. Complex number

A complex number z is a number which can be put in the form:

$$z = x + iy \tag{6.1}$$

Where, x and y are real numbers, and i is $\sqrt{-1}$. x is called a real part of the complex number z, and y is called the imaginary part. These are represented as follows:

$$R_e Z = x, I_m Z = y \tag{6-2}$$

Complex number and real number are generally classified into a group. There are solutions of equations which cannot be represented with real numbers. For example, the solution to the following equation cannot be found in the real number system:

$$x^2 + 2 = 0 \tag{6.3}$$

The solution to the above equation is $x = \pm\sqrt{-2}$. The square of a real number is 0 or positive (≥ 0). $\pm\sqrt{-2}$ does not satisfy this condition, and therefore, is not a real number. $\sqrt{-2}$ is represented as $2i$, by using the imaginary unit $i=+\sqrt{-1}$ for convenience' sake.

Complex number z can be represented geometrically with a point $P=(x,y)$ of the Cartesian coordinate system in Figure (7-1) and on a plane.

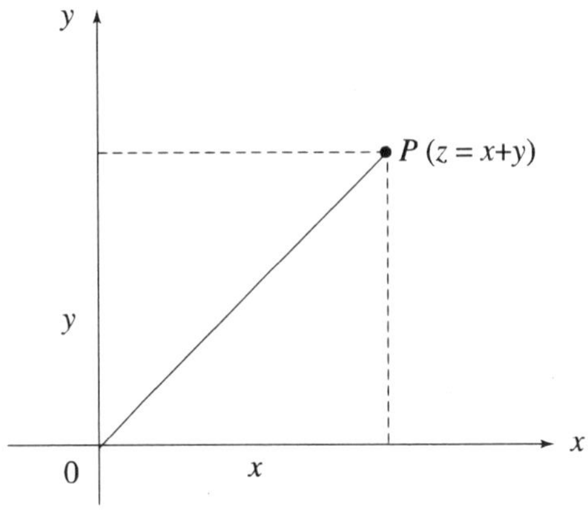

Figure 6-1 Geometric representation of complex number

The plane on which a complex number is marked is called a complex plane. The x-axis is the real (number) axis and the y-axis is the imaginary (number) axis.

1) Conjugate complex number

The complex conjugate of the complex number $z=x+iy$ is defined to be $\bar{z}=x-iy$. It is represented on the complex plane as in Figure 6-2.

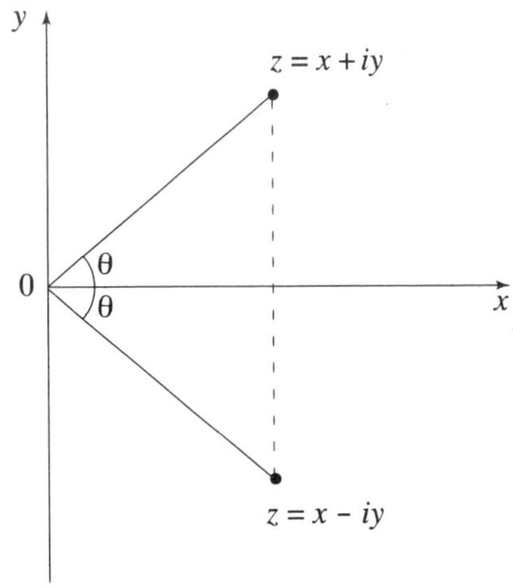

Figure 6-2 Geometric representation of conjugate complex number

2) Operation of complex number

Complex numbers z_1 and z_2 are as follows:

$$z_1 = x_1 + iy_1,\ \ z_2 = x_2 + iy_2 \tag{6-4}$$

The identity $z_1 = z_2$ means $x_1 = x_2,\ y_1 = y_2$.

The sum of complex numbers $z_1 + z_2$ is found by adding real numbers and imaginary numbers separately.

$$z_1 + z_2 = (x_1 + x_2) + i(y_1 + y_2) \tag{6-5}$$

The difference between complex numbers $z_1 - z_2$ is found with the following method:

$$z_1 - z_2 = (x_1 - x_2) + i(y_1 - y_2) \tag{6-6}$$

Figure 6-3 shows the addition and subtraction of complex numbers in a geometric way.

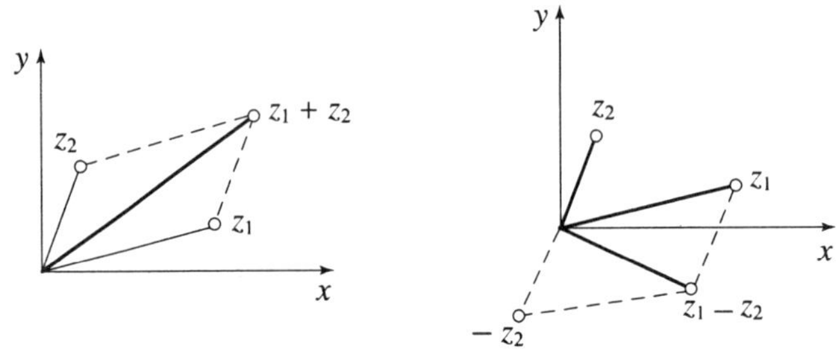

(a) sum (b) subtraction

Figure 6-3 Geometric representation of arithmetical operation

Multiplication of complex numbers $z_1 z_2$ is represented as follows:

$$z_1 z_2 = (x_1 + iy_1)(x_2 + iy_2) \tag{6-7}$$

The division of two complex numbers is defined by multiplying the denominator and numerator by the complex conjugate of the denominator.

$$\frac{z_1}{z_2} = \frac{x_1 + iy_1}{x_2 + iy_2} = \frac{(x_1 + iy_1)(x_2 - iy_2)}{(x_2 + iy_2)(x_2 - iy_2)}$$

$$= \frac{x_1 x_2 + y_1 y_2 + i(x_2 y_1 - x_1 y_2)}{{x_2}^2 + {y_2}^2}$$

$$= \frac{x_1 x_2 + y_1 y_2}{x_2^2 + {y_2}^2} + i\frac{x_2 y_1 - x_1 y_2}{{x_2}^2 + y_2} \tag{6-8}$$

The commutative law, associative law and distributive law are valid for operation of complex numbers.

① Commutative law: $z_1 + z_2 = z_2 + z_1,\ z_1 z_2 = z_2 z_1$

② Associative law: $(z_1 + z_2) + z_3 = z_1 + (z_2 + z_3),\ z_1(z_2 z_3) = (z_1 z_2) z_3$

③ Distributive law: $z_1(z_2 + z_3) = z_1 z_2 + z_1 z_3$

3) Properties of conjugate complex numbers

A complex number z and conjugate complex number $\bar{z}$ have the following properties:

First, addition makes a real number.

$$z+\bar{z}=2x \tag{6-9}$$

Second, multiplication makes a real number.

$$z\bar{z}=x^2+y^2 \tag{6-10}$$

Third, conjugation of addition, subtraction, multiplication and division of complex numbers z_1 and z_2 is the same with the conjugation of each complex number.

$$\overline{z_1+z_2}==\bar{z}_1+\bar{z}_2 \quad \overline{z_1-z_2}==\bar{z}_1-\bar{z}_2$$

$$\overline{z_1z_2}=\bar{z}_1\,\bar{z}_2,\ \overline{\left(\frac{z_1}{z_2}\right)}=\frac{\bar{z}_1}{\bar{z}_2} \tag{6-11}$$

Exercise 6-1

Find z_1z_2 and z_1/z_2 if $z_1=2-i2$ and $z_2=-3+i2$.

Solution

(1) $z_1z_2=(2-i2)(-3+i2)=-2+i10$

(2) $z_1/z_2=\dfrac{(2-i2)(-3-i2)}{(-3+i2)(-3-i2)}=\dfrac{-10+i2}{9+4}=-\dfrac{10}{13}+i\dfrac{2}{13}$

2. Polar form of complex number and De Moivre's theorem

The geometric representation of the complex number in Figure (6-1) can be represented using a vector with the magnitude (absolute value) of $|z|$ and the argument of θ as in Figure (6-4).

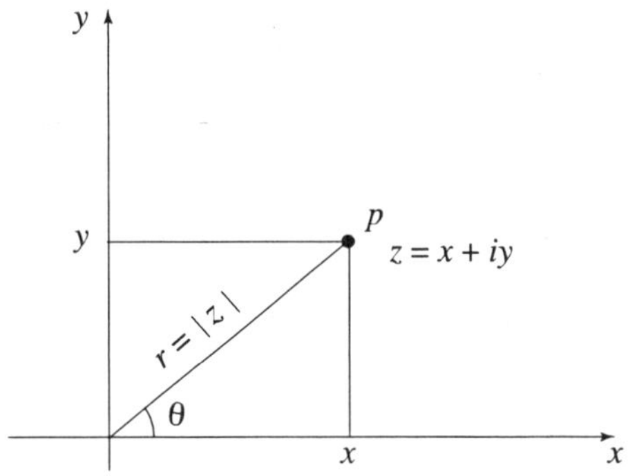

Figure 6-4 Vector representation of complex number

Where, the coordinates of P are as follows:

$$x = r\cos\theta, y = r\sin\theta \tag{6-12}$$

Therefore, the complex number z can be represented as follows:

$$z = x + iy = r(\cos\theta + i\sin\theta) \tag{6-13}$$

This is called the polar form of the complex number z, where, r is the distance from the origin to the point p, and the absolute value $|z|$ and argument θ of z are represented as follows:

$$r = |z| = \sqrt{x^2 + y^2}$$

$$\theta = \arg z = \tan^{-1}\frac{y}{x} \tag{6-14}$$

Exercise 6-2

Represent a complex number $z=1+i$ in a polar form.

Solution

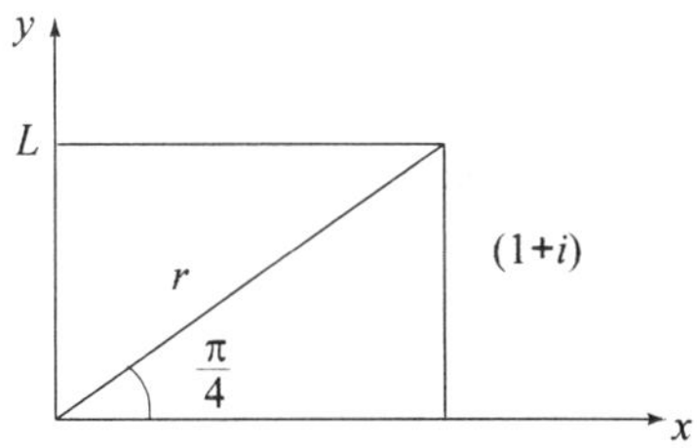

$$r=|z|=\sqrt{1^2+1^2}=\sqrt{2},\theta=\tan^{-1}\frac{1}{1}=\frac{\pi}{4}$$

$$\therefore z=1+i\sqrt{2}\left(\cos\frac{\pi}{4}+i\sin\frac{\pi}{4}\right)$$

Multiplication and division of two complex numbers can be more easily found in a polar form.

$$z_1=r_1(\cos\theta_1+i\sin\theta_1),z_2=r_2(\cos\theta_2+i\sin\theta_2) \quad (6\text{-}15)$$

$$\begin{aligned}z_1z_2&=r_1r_2(\cos\theta_1+i\sin\theta_1)(\cos\theta_2+i\sin\theta_2)\\&=r_1r_2\{(\cos\theta_1\cos\theta_2-\sin\theta_1\sin\theta_2)\\&\quad+i(\sin\theta_1\cos\theta_2-\cos\theta_1\sin\theta_2)\}\\&=r_1r_2\{\cos(\theta_1+\theta_2)+i\sin(Q_1+Q_2)\}\end{aligned} \quad (6\text{-}16)$$

$$\begin{aligned}\frac{z_1}{z_2}&=\frac{r_1}{r_2}\cdot\frac{\cos\theta_1+i\sin\theta_1}{\cos\theta_2+i\sin\theta_2}\\&=\frac{r_1}{r_2}\cdot\frac{\cos\theta_1+i\sin\theta_1}{\cos\theta_2+i\sin\theta_2}\cdot\frac{\cos\theta_2-i\sin\theta_2}{\cos\theta_2-i\sin\theta_2}\\&=\frac{r_1}{r_2}\cdot\{\cos(\theta_1-\theta_2)+i\sin(\theta_1-\theta_2)\}\end{aligned} \quad (6\text{-}17)$$

Therefore, if the argument of two complex numbers are known, we can perform multiplication and division of two complex numbers by simply adding or subtracting the two arguments. Because the trigonometric function part of (6-16) and (6-17) is 1, the absolute value of the multiplication of two complex numbers equals the multiplication of absolute values of each complex number, and the argument equals the sum of the amplitudes of each complex number.

$$\begin{aligned} |z_1 z_2| &= |z_1|\,|z_2|, \theta = \theta_1 + \theta_2 \\ &= r_1 r_2 \angle \theta \end{aligned} \tag{6-18}$$

In the same way, the absolute value of division of two complex numbers is as seen in (6-19).

$$\begin{aligned} \left|\frac{z_1}{z_2}\right| &= \frac{|z_1|}{|z_2|}, \theta = \theta_1 - \theta_2 \\ &= \frac{r_1}{r_2} \angle \theta \end{aligned} \tag{6-19}$$

Extending (6-16) and evaluating z^n is represented as follows:

$$\begin{aligned} z^n &= r^n (\cos\theta + i\sin\theta)^n \\ &= r^n (\cos n\theta + i\sin n\theta) \end{aligned} \tag{6-20}$$

This formula is called De Moivre's formula.

Exercise 6-3

Find z_1/z_2 for two complex numbers z_1 and z_2.

Solution

$$z_1 = 2\left(\cos\frac{\pi}{3} + i\sin\frac{\pi}{3}\right),\ z_2 = \left(\cos\frac{\pi}{6} + i\sin\frac{\pi}{6}\right)$$

$$z_1 z_2 = 2\times 1\left(\cos\frac{\pi}{2} + \sin\frac{\pi}{2}\right).$$

$$= i\,2$$

$$z_1/z_2 = \frac{2}{1}\left(\cos\frac{\pi}{6} + i\sin\frac{\pi}{6}\right)$$

$$= \sqrt{3} + i$$

Exercise 6-4

Find $(1+i\sqrt{3})^2$ by using De Moivre's formula.

Solution

$$(1+i\sqrt{3}) = 2\left(\cos\frac{\pi}{3} + i\sin\frac{\pi}{3}\right)$$

$$(1+i\sqrt{3})^2 = 2^2\left(\cos\frac{2}{3}\pi + i\sin\frac{2}{3}\pi\right)$$

$$= -2 + i2\sqrt{3}$$

In the triangle inequality of the absolute values of two complex numbers, the absolute value of the sum of two complex numbers is equal to or smaller than the sum of the absolute values of the complex numbers.

$$|z_1 + z_2| \leq |z_1| + |z_2| \tag{6-21}$$

3. Analytic function of complex number

1) Analytic function

Let's define the function of a complex number. It is defined using the same concepts as a normal function. Assume two complex numbers z and w. Each value on the z plane corresponds to a value on the w plane. In this case, w is the function of z as defined in the area, and is represented as $w=f(z)$. If the complex number z changes in the area, the set of complex numbers available for w is called the range of values of the function $w=f(z)$. Because w is subordinate to z, it can be represented as the following:

$$w=u(x,y)+iv(x,y) \tag{6-22}$$

Exercise 6-5

When $w=f(z)=z^2+3z$, mark the real number and imaginary number of w on the z plane.

Solution

$u(x,y)=R_e f(z)=x^2-y^2+3x$

$dv(x,y)=I_m f(z)=3xy+3y$

Let's find the limit, continuity and differentiability of a function on a complex number plane.

In Figure (6-5), if the function $f(z)$ is defined near z_0, and a positive real number δ is evaluated with the following formula (6-23) for all the values $z(z \neq z_0)$that satisfies $|z-z_0|<\delta$ for a positive real number ϵ, the limit is $l$when z approaches to z_0:

$$|f(z)-l|<\epsilon \tag{6-23}$$

It is written as follows:

$$\lim_{z \to z_0} f(z) = l \tag{6-24}$$

The value is consistent when z approaches to z_0 from any direction on the complex number plane.

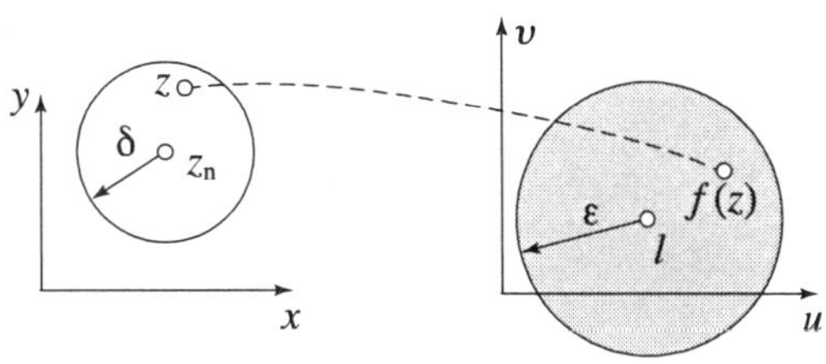

Figure 6-5 Limit

If $f(z_0)$ is defined and satisfies $\lim_{z \to z_0} f(z) = f(z_0)$, $f(z)$ is continuous in $z=z_0$. If the following is the only limit, it is differentiable in $z=z_0$.

$$f'(z_0) = \lim_{\Delta_z \to 0} \frac{f(z_0+\Delta z)-f(z_0)}{z-z_0} \tag{6-25}$$

In the formula (6-25), $f'(z_0)$ is the derivatives of $f(z)$ in $z=z_0$.

$$f'(z) = \lim_{z \to z_0} \frac{f(z)-f(z_0)}{z-z_0} \tag{6-26}$$

The value is constant regardless of the direction in which z approaches z_0.

Exercise 6-6

The derivatives of function $f(z)=z^2$ is $2z$ in z. Prove that there is no derivatives in a point in $f(z)=\bar{z}=x-iy$.

Solution

The derivatives of z^2 is $f'(z_0)=\lim\limits_{\Delta_z \to 0}\dfrac{(z+\Delta z)^2-z^2}{z-z_0}=\lim\limits_{\Delta_z \to 0}=2z$.

Where, the derivatives of $\bar{z}$ is as follows:

$$f'(z)=\lim_{\Delta_z \to 0}\frac{f(z+\Delta z)-f(z)}{z-z_0}=\lim_{\Delta_x,\Delta_y \to 0}\frac{(x+\Delta x)-i(y+\Delta y)-(x-iy)}{\Delta x+i\Delta y}$$

$$=\lim_{\Delta_x,\Delta_y \to 0}\frac{\Delta x-i\Delta y}{\Delta x+i\Delta y}$$

The value of this formula varies depending on the order of Δx and Δy approaching 0, and thus, there are no derivatives. In other words, if $\Delta x \to 0$ after $\Delta y \to 0$, the limit approaches 1. If $\Delta y \to 0$ after $\Delta x \to 0$, the limit approaches -1.

2) Exponential function

For complex number $z=x+iy$, the exponential function e^z can be represented as the following:

$$e^z=e^{x+iy}=e^x(\cos y+i\sin y) \tag{6-27}$$

Where, e^{iy} is as the following:

$$e^{iy}=\cos y+i\sin y \tag{6-28}$$

This formula is called Euler's formula.

The characteristics of the exponential function of a complex number can be applied in the same way as those of real numbers. Formula (6-29) shows various formulas for the exponential function e^x of real number.

$$(e^x)' = e^x$$

$$e^{x+x_2} = e^{x_1} \cdot e^{x_2}$$

$$e^x = 1 - x + \frac{x^2}{2!} + \frac{x^3}{3!} + \cdots \quad \text{(Maclaurin's series)} \tag{6-29}$$

Various formulas are derived for the exponential function e^2 of complex number. The formula for derivatives $(e^2)'$ is as follows:

$$(e^z)' = \frac{\partial}{\partial x}(e^x \cdot \cos y) + i\frac{\partial}{\partial x}(e^x \cdot \sin y)$$

$$= e^x \cdot \cos y + i\, e^x \cdot \sin y \tag{6-30}$$

It is the same as e^z of Formula (6-27). With $z_1 = x_1 + iy_1, z_2 = x_2 + iy_2$, the following is made, using the addition theorem of sine and cosine:

$$e^{z_1 + z_2} = e^{z_1} \cdot e^{z_2} \tag{6-31}$$

The following formulas can be proven with Euler's formula:

$$z = r(\cos\theta + i\sin\theta) = re^{i\theta}$$

$$|e^{iy}| = \sqrt{\cos^2 y + \sin^2 y} = 1$$

$$|e^z| = e^x$$

$$e^{2\pi i} = 1, \quad e^{\pi} = e^{-\pi i} = -1$$

$$e^{\pi i/2} = i, \quad e^{-\pi i/2} = -i \tag{6-32}$$

e^z is a periodic function with the imaginary number period of $2\pi i$.

$$e^{z \pm 2\pi i} = e^x, \quad n = 0, 1, 2, \cdots \tag{6-33}$$

Because of this periodic feature, in the equation $w = e^z$, the value of z can be within a strip ranging from $-\pi$ to π on the y-axis in the complex plane.

Exercise 6-7

Evaluate e^z when $z=2+i\frac{\pi}{2}$.

Solution

$$e^z=e^2\cdot e^{i\frac{\pi}{2}}=e^2\left(\cos\frac{\pi}{2}+i\sin\frac{\pi}{2}\right)$$

$$=ie^2$$

3) Trigonometric function

Using Euler's formula for exponential functions, we can find various properties of trigonometric function.

$$e^{ix}=\cos x+i\sin x,\quad e^{-ix}=\cos x-i\sin x \tag{6-34}$$

In (6-34), $\cos x$ and $\sin x$ are as follows:

$$\begin{cases}\cos x=\dfrac{1}{2}(e^{ix}+e^{-ix})\\ \sin x=\dfrac{1}{2i}(e^{ix}-e^{-ix})\end{cases} \tag{6-35}$$

This also applies to the complex number $z=x+iy$. A trigonometric function has the following various properties:

$$\begin{cases}\cos z=\dfrac{1}{2}(e^{iz}+e^{-iz})\\ \sin z=\dfrac{1}{2i}(e^{iz}-e^{-iz})\end{cases} \tag{6-36}$$

$$\tan z=\frac{\sin z}{\cos z},\ \cot z=\frac{\cos z}{\sin z}$$

$$\sec z=\frac{1}{\cos z},\ \csc z=\frac{1}{\sin z} \tag{6-37}$$

Because $\cos z$ and $\sec z$ are even functions, and the others are odd functions, the followings are valid:

$$\cos(-z) = \cos z, \qquad \sin(-z) = -\sin z$$
$$\cot(-z = -\cot z, \qquad \tan(-z) = -\tan z$$
$$\cos(z \pm 2n\pi) = \cot z, \qquad \sin(z \pm 2n\pi) = \sin z$$
$$\cot(z \pm n\pi) = \cot z, \qquad \tan(z \pm n\pi) = \tan z$$
$$n = 0,1,2,\cdots \tag{6-38}$$

From Formula (6-36), the following formulas are made:

$$\frac{d}{dz}cosz = -\sin z, \frac{d}{dz}sinz = \cos z, \quad \frac{d}{dz}tanz = \sec^2 z$$
$$\cos(z_1 \pm z_2) = \cos z_1 \cdot \cos z_1 \cdot \cos z_2 \mp \sin z_1 \cdot \sin z_2$$
$$\sin(z_1 \pm z_2) = \sin z_1 \cdot \cos z_1 \cdot \cos z_2 \pm \sin z_2 \cdot \cos z_1$$
$$\sin^2 z + \cos^2 z = 1 \tag{6-39}$$

Euler's formula $e^{iz} = \cos z + i\sin z$ for complex number is valid. The following formulas are made from the $\cos(z_1 \pm z_2)$ formula:

$$\cos(x+iy) = \cos x \cdot \cos iy - \sin x \cdot \sin iy$$
$$\sin(x+iy) = \sin x \cdot \cos iy + \sin x \cdot \cos iy \tag{6-40}$$

4) Hyperbolic function

The hyperbolic function of a complex number is defined by the following in the case of a real numbers:

$$\sinh z = \frac{e^z - e^{-z}}{2}, \cosh z = \frac{e^z + e^{-2}}{2} \tag{6-41}$$

Tanhz, cothz, sechz and cosechz are similar to Formula (6-37).

Derivatives of a hyperbolic function are also similar to (6-39). The following functions are widely used:

$$\sinh(-z) = -\sinh z$$

$$\cosh(-z) = \cosh z$$

$$\sinh^2 z - \cosh^2 z = 1$$

$$\sinh(z_1 + z_2) = \sinh z_1 \cdot \cosh z_2 + \cosh z_1 \cdot \sinh z_2$$

$$\cosh(z_1 + z_2) = \cosh z_1 \cdot \cosh z_2 + \sinh z_1 \cdot \sinh z_2$$

$$\cosh(x + iy) = \cosh x \cdot \cos y + i \sinh x \cdot \sin y$$

$$\sinh(x + iy) = \sinh x \cdot \cos y + i \cosh x \cdot \sin y \qquad (6\text{-}42)$$

4. Differentiation and integration on complex planes

Let's find the expression of the area of complex functions first. If the distance between a variable z and a constant a is a circle with the center a and radius ρ, it can be represented as Formula (6-34).

$$|z - a| = \rho \qquad (6\text{-}43)$$

For an inequation (6-44), the area can be represented as in Figure (6-6)(b). Therefore, $|z| = 1$ indicates the unit element as in Figure (6-6)(c).

$$\rho_1 < |z - a| < \rho_2 \qquad (6\text{-}44)$$

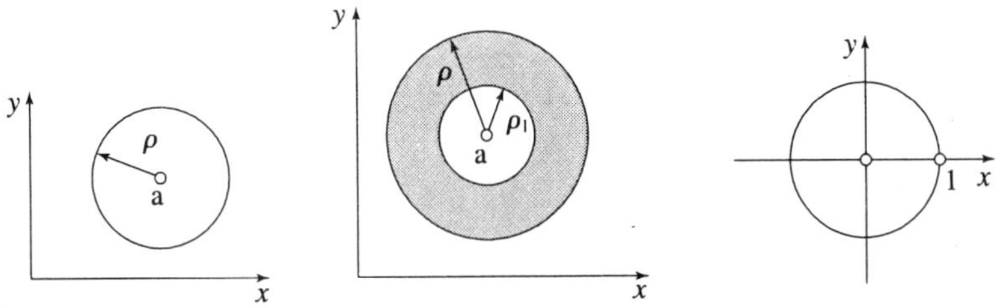

(a) $|z-a|=\rho$ (b) $\rho_1 < |Iz-a| < \rho_2$ (a) $|z|=l$

Figure 6-6 Area of complex function

1) Differentiation of complex number

$$w=f(z)=u(z,y)+iv(x,y) \tag{6-45}$$

In the above complex function, if the function $f(z)$ is defined at all the points around z_0 and has the derivatives $f'(x)$, it is analytic in $z=z_0$.

$$f'(z)=\lim_{\Delta_z\to 0}\frac{f(z+\Delta z)-f(z)}{\Delta z} \tag{6-46}$$

In (6-46), $f'(z)$ must have the same value regardless of the path on which Δz approaches 0. As $\Delta z=\Delta X+i\Delta y$, let's evaluate derivatives for paths I and II as in Figure (6-7).

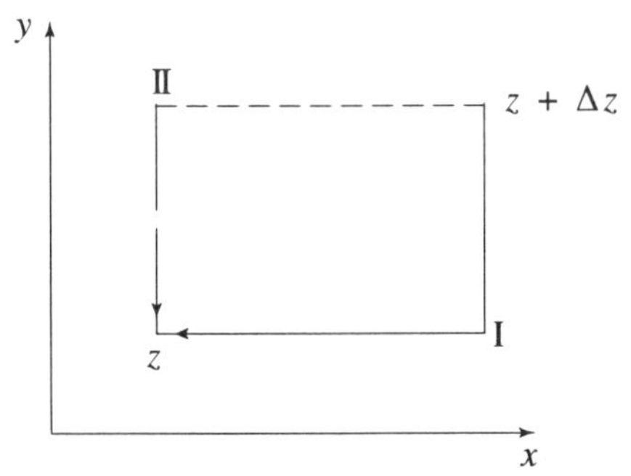

Figure 6-7 Access paths of (6-46)

When Δz approaches 0 on Path I, with $\Delta y \to 0$ before $\Delta x \to 0$. If $\Delta y \to 0$ is performed first in $\Delta z = i\Delta y$, then, $\Delta z = \Delta x$, and $f'(z)$ is as follows, inputting (6-45) into (6-46):

$$f'(z) = \lim_{\Delta x \to 0} \frac{[u(x+\Delta x, y) + iv(x+\Delta x, y)] - [u(x,y) + iv(x,y)]}{\Delta x}$$

$$= \lim_{\Delta x \to 0} \frac{u(x+\Delta x, y) - u(x,y)}{\Delta x} + i \lim_{\Delta z \to 0} \frac{v(x+\Delta x, y) - v(x,y)}{\Delta x}$$

$$= \frac{\partial u}{\partial x} + i \frac{\partial v}{\partial x} \quad (6\text{-}47)$$

Similarly, if $\Delta x \to 0$ is performed first on Path II, then, $\Delta z = i\Delta y$, and $f'(z)$ is as follows:

$$f'(z) = \lim_{\Delta y \to 0} \frac{u(x, y+\Delta y) - u(x,y)}{i\Delta y} + i \lim_{\Delta y \to 0} \frac{v(x, y+\Delta y) - v(x,y)}{i\Delta y}$$

$$= \frac{\partial v}{\partial y} - i \frac{\partial u}{\partial y} \quad (6\text{-}48)$$

Because the derivatives formula (6-47) for Paths I and II must be the same with (6-48), the real number equals the imaginary number.

$$\frac{\partial u}{\partial x} = \frac{\partial v}{\partial y}$$

$$\frac{\partial u}{\partial y}=\frac{\partial v}{\partial x} \tag{6-49}$$

This formula is a condition for the existence of differentiation of function, and is called Cauchy- Riemann's differential equation.

Exercise 6-8

Prove that $f(z)-z^2=(x^2-y^2)+i(2xy)$ is an analytic function for every z.

Solution

Because $u=x^2-y^2, v=2xy$,

$$\frac{\partial u}{\partial x}=2x,\quad \frac{\partial v}{\partial x}=-2y,$$

$$\frac{\partial u}{\partial x}=2x,\quad \frac{\partial v}{\partial y}=2x$$

It is an analytic function since it satisfies the Cauchy-Riemann equation.

2) Integration of complex number

Line integral $\int_a^b g(x)dx$ on a complex plane can be converted to an integration formula $\int_f (z)dz$ for the complex function following a curve c on the complex plane. In Figure (6-8), if two real number functions $x(t)$ and $y(t)$ for time t are given, the points $(x(t), y(t))$ on the complex plane moves following the curve c until t changes from a to b. If the curve c on the z plane is differentiable at each point of the curve, it is represented as the following formula (6-50):

$$z(t)=x(t)+iy(t)\ (a \leqq t \leqq b) \tag{6-50}$$

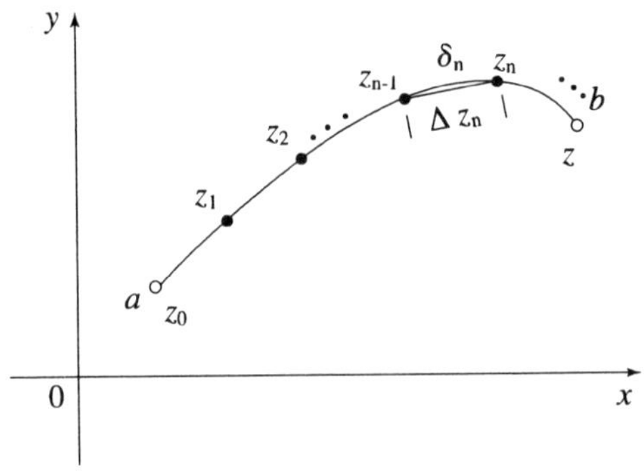

Figure 6-8 Line integration on a complex plane

As in Figure (6-8), divide the path c into $z_1, z_2, \cdots, z_{n-1}, z_n$, and multiply z_{n-1} by z_n for the entire section.

$$\sum_{m=1}^{n}(\delta_m)\Delta z = \int_c f(z)dz \tag{6-51}$$

Formula (6-51) is the line integral of $f(z)$ which follows path C.

As $f(z) = u(x,y) + iv(x,y)$ and $dz = dx + idy$, (6-51) can be written as follows:

$$\int_c f(z)dz = \int_c f(u+iv)(dx+idy)$$
$$= \int_c udx - \int_c vdy + i\left[\int_c udy + \int_c vdx RIGHT\right] \tag{6-52}$$

If the integration path c is given as a function of time t as in (6-50), it is calculated as follows:

$$\int_c (z)dz = \int_a^b u \cdot \dot{x}dt - \int_a^b v \cdot \dot{y}dt + i\left[\int_c udy + \int_c vdx\right] \tag{5-53}$$

Where, $\dot{x}$ and $\dot{y}$ are integration for time of x and y, and $u = u[x(t), y(t)]$ and $v = v[x(t), y(t)]$. Formula (6-53) can be simplified to the following:

$$\int_c f(z)dz = \int_a^b (u+iv)(\dot{x}+i\dot{y})dt$$

$$= \int_a^b f[z(t)]\dot{z}(t)dt \tag{6-54}$$

Exercise 6-9

Integrate $f(z) = \frac{1}{z}$ and $f(z) = \frac{1}{z^2}$ as Path c starting at $z=1$ turns around a unit circle counter-clockwise.

Solution

The path c with the unit circle $|z| = 1$ is represented as the following:

$z(t) = e^{it} = \cos t + i\sin t (0 \leq t \leq 2\pi)$

$\dot{z}(t) = ie^{it} = -\sin t + i\cos t$

$$\int_c \frac{1}{z} dz = \int_o^{2\pi} \frac{1}{e^{it}} \cdot ie^{it}dt = i\int_0^{2\pi} dt = 2\pi i$$

$$\int_c \frac{1}{z^2} dz = \int_o^{2\pi} \frac{1}{(e^{it^2}} \cdot ie^{it}dt = \int_0^{2\pi} ie^{-it}$$

$$= i\int_0^{2\pi} \cos t dt + \int_0^{2\pi} \sin t dt = 0$$

Exercise 6-10

Integrate $f(z) = R_e z = x$ following c_1 or c_2 as the path c starts at $z=0$ and moves to $z=1+i$.

Solution

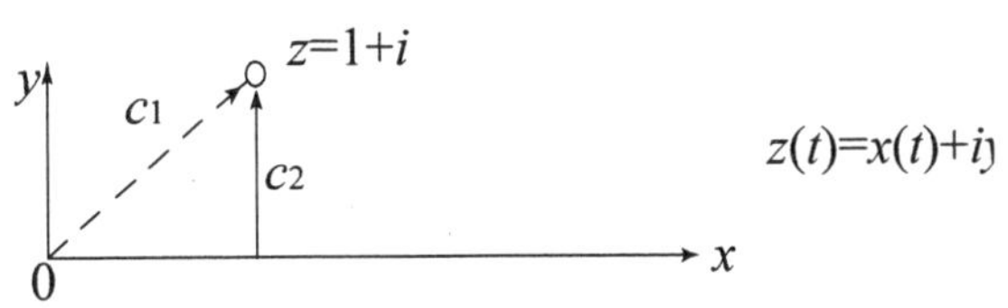

When integrating through the path c_1, the equation of c_1 is $z(t) = (1+i)t$. Therefore, $R_e z(t)$, $z(t) = (1+i)t$, and the following is made.

$$\int_c R_e z dz = \int_0^1 t(1+i)dt = (1+i)\int_0^1 tdt$$

$$= \frac{1}{2}(1+i)$$

When integrating through the path C_2, it is as follows:

$$\int_c R_e z dz = \int_0^1 tdt + \int_1^2 idt = \frac{1}{2} + i$$

Because the $R_e z$ integration value varies between c_1 and c_2, $f(z) = x$ is not analytic.

Cauchy's theorem

If the function $f(z)$ is analytic in the closed curve, bounded domain D, the following formula is valid for each simply connected closed curve c in the area D of Figure (6-9):

$$\int_c f(z)dz = 0 \qquad (6\text{-}55)$$

This important theorem is called Cauchy's theorem. It can be proven as follows:

$$\int_c f(z)dz = \int_c (udx - vdy) + i\int_c (udy + vdx) \qquad (6\text{-}56)$$

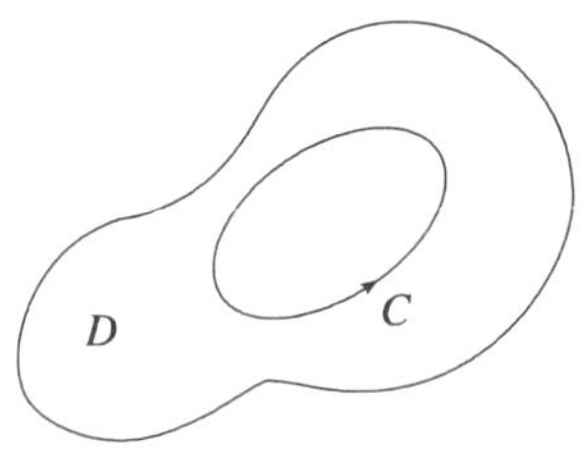

Figure 6-9 Cauchy's theorem

In (6-56), $f(z)$ is an analytic function, and if there are derivatives, it is possible to 1st-order differentiate variables u and v in area D. Applying Green's theorem produces the following formula:

$$\int_c (udx - vdy) = \int_R\int\left(-\frac{dv}{dx} - \frac{\partial u}{\partial y}\right)dxdy \tag{6-57}$$

Where R is the area enclosed by the closed curve c. Likewise, the following formula is as valid:

$$\int_c (udx + vdy) = \int_R\int\left(\frac{\partial v}{\partial x} - \frac{\partial u}{\partial x}\right)dxdy \tag{6-58}$$

If Cauchy-Riemann's equation (6-49) is applied to (6-57) and (6-58), the rights sides of the two formulas become 0. Formula (6-56) becomes 0, and therefore, Cauchy's theorem is valid. Let's integrate a simple closed curve with two integration paths as seen in Figure (6-10).

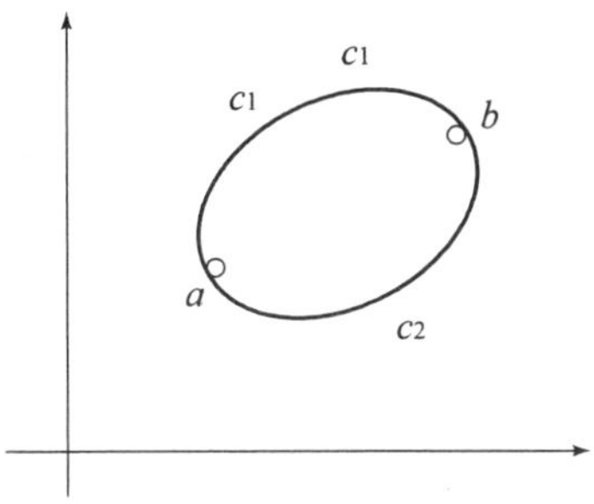

Figure 6-10 Simple closed curve

In Cauchy's theorem $\int_c f(z)dz = 0$,

$$\int_{c_2} f(z)dz = \int_{c_1} f(z)dz + \int_{c_2} f(z)dz$$

$$= \int_{c_2} f(z)dz - \int_{c_1} f(z)dz$$

$$=0 \tag{6-59}$$

If the sectional integration path is opposite, the negative symbol is applied. Consider integration of $\int_c \frac{1}{z}dz$ as mentioned in Exercise 9. Even though $\frac{1}{z}$ is analytic on the integration path c, the integration value is not 0 as Cauchy's theorem would have us expected. This is because the function is not analytic if the area D includes $z=0$. Integration of $\int_c \frac{1}{z^2}dz$ under the same conditions is 0, and thus this theorem defies simple explanation.

However, Cauchy's theorem is an important foundation for the analysis of complex number. If we define the area D excluding the non-analytic points, as shown in Figure (6-11), the closed curve is not simple-connected but multi-connected, and it is impossible to apply Cauchy's theorem. To solve this problem, we can make a very narrow connection c_3 in area D to create a simple closed curve $c_3c_1c_3c_2$, and apply Cauchy's theorem. If $f(z)$ is analytic on c_1c_2 and between the points, the following Formula (6-60) is valid.

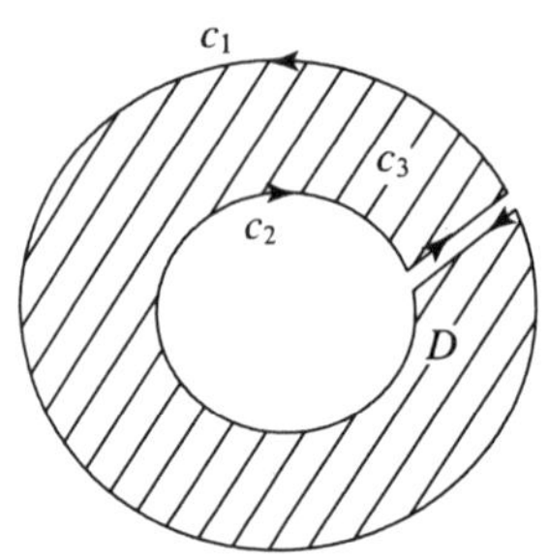

Figure 6-11 Application of Cauchy's theorem to multiple closed curve

$$\int_{c_1} f(z)dz + \int_{c_2} f(z)dz \int_{L_1} f(z)dz + \int_{L_2} f(z)dz = 0 \tag{6-60}$$

Path L is 0 because it is integrated in the opposite direction.

This is an application of Cauchy's theorem, and is used to integrate real numbers on complex planes.

Exercise 6-11

c_1 is a unit circle $|z| = 11$, and c_2 is a circle $|z| = \frac{1}{2}$, integrate $f(z) = \frac{1}{z}$ for two paths.

Solution

Using the result of Exercise 9, it is found that $\int_{c_1} \frac{1}{z} dz = 2\pi i$, provided that c_1 and c_2 are in the same direction.

Cauchy's integration formula

For Figure (6-11), the application of Cauchy's theorem may be extended to derive a new integration formula. As in Figure (6-12), if $f(z)$ is analytic in the simple-connected area D, for point z_0 in D and circle c_2 with the radius R enclosing points z_0, the following formula is valid.

$$\int_{c_1} \frac{f(z)}{z - z_0} dz = 2\pi i f(z_0) \tag{6-61}$$

This is called Cauchy's integration formula.

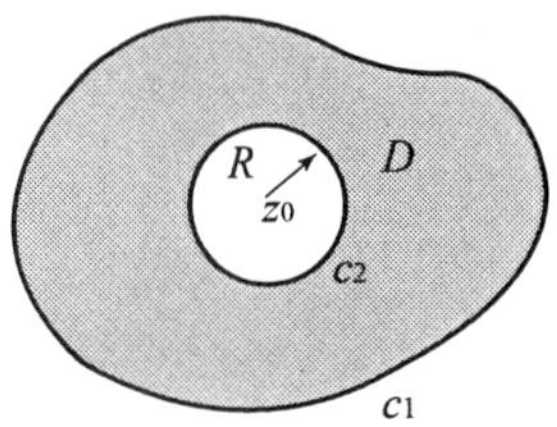

Figure 6-12 Cauchy's integration formula

Let's prove (6-61). Substituting $f(z)=f(z_0)+[f(z_0)-f(z_0)]$ for (6-61) makes the following formula:

$$\int_c \frac{f(z)}{z-z_0}dz = f(z_0)\int_c \frac{dz}{z-z_0} + \int_c \frac{f(z)-f(z_0)}{z-z_0}dz \tag{6-62}$$

In (6-62), the first member is $2\pi i f(z_0)$. If we calculate the second member and reduce the radius R substantially in Figure (6-12), the following formula is valid for point z on c_2:

$$| f(z)-f(z_0) | < \epsilon, \ | z-z_0 | = R \tag{6-63}$$

Where ϵ is a sufficiently small value. (6-63) becomes the following:

$$| f(z)-f(z_0) | / | z-z_0 | < \frac{\epsilon}{R} \tag{6-64}$$

For $a=0$, the formula expands to (6-69).

$$f(z)=f(0)+f'(0)z+\frac{f''(0)}{2!}z^2+...+\frac{f^n(0)}{n!}z^n+... \tag{6-69}$$

This is called Maclaurin's expansion. It is a Taylor series with the center a of 0.

Exercise 6-12

Evaluate Maclaurin series of $f(z)=\frac{1}{1-z}$.

Solution

nth derivatives is $f^n(z)=\frac{n!}{(1-z)^{n-1}}$, and therefore, $f^n(0)=n!$.

$$\frac{1}{z-1}=\sum_{n=0}^{\infty} z^n = 1+z+z^n+\cdots \ (| z | <1) \tag{6-70}$$

Exercise 6-13

Evaluate Maclaurin series of $f(z)=e^{x}$, and evaluate the expansion of sin z and cos z, using Euler's formula.

Solution

e^z is analytic for every z, and $(e^z)'=e^z$, therefore, the following is made.

$$e^z=\sum_{n=0}^{\infty}\frac{z^n}{n!}=1+z+\frac{z^2}{2!}+\cdots$$

By substituting $z=iy$ and using Euler formula $e^{iy}=\cos y+i\sin y$, we can evaluate Maclaurin series of real functions $\cos y$ and $\sin y$.

$$e^{iy}=\sum_{n=0}^{\infty}\frac{(iy)^n}{n!}=\sum_{n=0}^{\infty}(-1)^k\frac{y^{2k}}{(2k)!}+i\sum_{k=0}^{\infty}(-1)^k\frac{y^{2k+1}}{(2k+1)!}$$
$$=\cos y+i\sin y$$

where,

$$\cos y=\sum_{k=0}^{\infty}(-1)^k\frac{y^{2k}}{(2k)!}=1-\frac{y^2}{2!}+\frac{y^4}{4!}-\cdots$$

$$\sin y=\sum_{k=0}^{\infty}(-1)^k\frac{y^{2k+1}}{(2k+1)!}=y-\frac{y^3}{3!}+\frac{y^5}{5!}-\cdots$$

3) Laurent series

If a function $f(z)$ is not defined in a point a, there are cases that the range of area D should be developed in $Z=a$. Because no value is defined in $Z=1$in Exercise 13, the function is not analytic, and the range of $z\geqq 1$ with $Z=1$ as the boundary cannot be developed as a Taylor series. The $z=1$ point is called a singular point of $f(z)$. A Laurent series expansion, which is related with the singular point where analyticity is lost, is an important base for application of complex integration, such as a singular point, residue theorem and the calculation of real number integration.

If $f(z)$ is not analytic at point a, and as in Figure (6-14), $f(z)$ is analytic in the ring of two concrete circles c_1, c_2 with the center a, $f(z)$ can be developed as the following Laurent series:

$$f(z) = \sum_{n=0}^{\infty} b_n (z-a)^n \frac{c_n}{(z-a)^n} \qquad (6\text{-}71)$$

Provided that,

$$b_n \frac{1}{2\pi i} \int_c \frac{f(z^*)}{(z^* - a)^{n-1}} dz^*$$

$$c_n \frac{1}{2\pi i} \int_c (z^* - a)^{n-1} f(z^*) dz^*$$

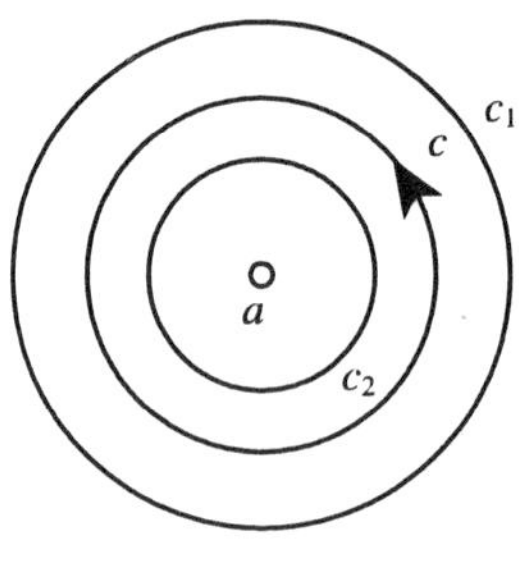

Figure 6-13 Laurent theorem

In the formula (6-71), if $f(z)$ is analytic at a, and is a Taylor series, the coefficient c_n becomes 0. In reality, there is hardly a case that we should evaluate Laurent series expansion by calculating integral equation. Look at the following exercises.

Exercise 6-14

Evaluate Laurent series of $f(z)\frac{e^{2x}}{(z-1)^3}$ at a given singular point $z=1$.

Solution

$z-1=u$, and $z=1+u$

$$f(z)=\frac{e^{2z}}{(z-1)^3}=\frac{e^{2+2u}}{u^3}=\frac{e^2}{u^3}\cdot e^{2u}$$

Laurent series expansion is normally evaluated from Taylor series expansion. Therefore, e^{2u} is expanded to Taylor series as the following:

$$e^{2u}=1+2u+\frac{(2u)^2}{2!}+\frac{(2u)^3}{3!}+\cdots$$

Therefore,

$$f(z)=\frac{e^2}{u^3}\left\{1+2u+\frac{(2u)^2}{2!}+\frac{(2u)^3}{3!}+\cdots\right\}$$

$$=\frac{e^2}{(z-1)^3}+\frac{2e^2}{(z-1)^2}+\frac{2e^2}{(z-1)}+\frac{4e^2}{3}+\frac{2e^2}{3}(z-1)+\cdots$$

This series is converted for every z of $z\neq 1$.

4) Residue

If $f(z)$ is analytic near $z=a$, $\int_c f(z)dz=0$ is valid by Cauchy's theorem. If $f(z)$ becomes a singular point at $z=a$, Cauchy's theorem is not generally valid . In this case, it can be represented with Laurent series so that convergence is made at $0<|z-a|<R$. If $n=1$ in the latter member of this series, the coefficient c_1 of $\frac{1}{z-a}$ is as follows:

$$c_1=\frac{1}{2\pi i}\int_c f(z)dz,\ \int_c f(z)dz=2\pi i\cdot c_1 \quad (6\text{-}72)$$

Coefficient c_1 is called a residue in $z=a$ of $f(z)$, and is represented with the following symbol:

$$c_1 = \underset{z=a}{Res} f(z) \tag{6-73}$$

Fromhere we can find a simple way to determine the residue.
If $f(z)$ is a simple pole at $z=a$, the corresponding Laurent series is as follows:

$$f(z) = \frac{c_1}{z-a} + b_0 + b_1(z-a) + b_2(z-a)^2 + \cdots \tag{6-74}$$

Multiplying both sides of (6-73) by $z-a$ and having z approach a will make the right side approach c_1.

$$\underset{z=a}{Res} f(z) = c_1 \lim_{z \to a}(z-a)f(z) \tag{6-75}$$

If $f(z)$ has the pole of m degree $(m>1)$ at $z=a$, it is represented as the following:

$$f(z) = \frac{c_m}{(z-a)^n} + \frac{c_{m-1}}{(z-a)^{m-1}} + \cdots + \frac{c_2}{(z-a)^2} + \frac{c_1}{(z-a)^1}$$
$$+ b_0 + b_1(z-a) + \cdots \tag{7-76}$$

Multiplying both sides by $(z-a)^m$ makes the following formula:

$$f(z)(z-a)^m = c_m + c_{m-1}(z-a) + \cdots + c_2(z-a)^{m-2} + c_1(z-a)^{m-1}$$
$$+ b_0(z-a)^m + b_1(z-a)^{m+1} + \cdots \tag{6-77}$$

This formula shows that residue c_1 at $z=a$ of $f(z)$ is a coefficient of $(z-a)^{m-1}$ in the Taylor series of $g(z)=(z-a)^m f(z)$. Therefore, the residue is represented as the following by Taylor's theorem:

$$c_1 = \frac{g^{m-1}(a)}{(m-1)!} \tag{6-78}$$

The residue is derived as follows:

$$\underset{z=a}{Res} f(z) = \frac{1}{(n-1)} \lim_{z \to a} \left\{ \frac{d^{m-1}}{dz^{m-1}} [(z-a)^m f(a)] \right\} \tag{6-79}$$

Consider Figure (6-15) where $f(z)$ has a finite number of singular points $a_1, a_2, \cdots a_n$ in a simple closed curve c. If the function is analytic inside except on singular points and on c, the following formula is made by Cauchy's integral theorem:

$$\int_c f(z)dz = \int_{c_1} f(z)dz + \int_{c_2} f(z)dz + \cdots + \int_{c_m} f(z)dz \tag{6-80}$$

$$\int_{c_1} f(z)dz = 2\pi i \underset{z=a}{Res} f(z), j = 1, 2, \cdots, m \tag{6-81}$$

This is called a residue theorem. In the integration path, as in Figure (6-15), the integration is made from c to c_1, from c_1 to c_2, up to c_m, and returned to the last c along the narrow path L.

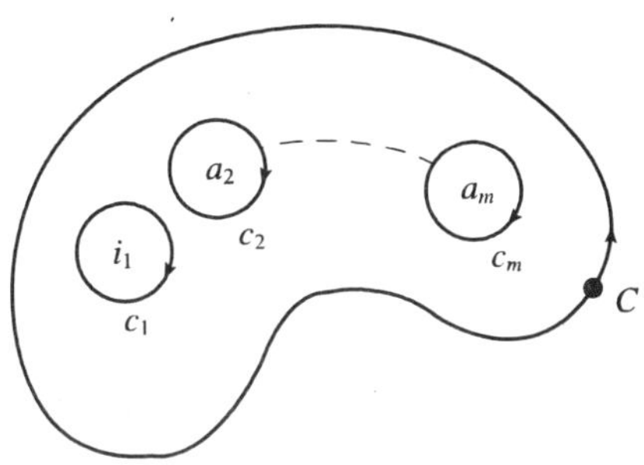

Figure 6-14 Residue theorem

Exercise 6-15

Integrate $\frac{1}{(z-1)^2(z-3)}$ along the path c of the circumference $|z-1|=1$.

Solution

This function has a secondary pole at $z=1$ and a single pole at $z=3$. Because only $z=1$ is included in the path c, according to (6-81), the following is made:

$$\int_c \frac{dz}{(z-1)^2(z-3)} = 2\pi i \underset{z=a}{Res} \frac{1}{(z-1)^2(z-3)}$$

$$= 2\pi i \cdot \frac{1}{(2-1)!} \lim_{z\to 1}\left\{\frac{d}{dz}\left[(z-1)^2 \frac{1}{(z-1)^2(z-3)}\right]\right\}$$

$$= 2\pi i \lim_{z\to 1}\frac{-1}{(z-3)^2}$$

$$= -\frac{\pi}{2} i$$

In Exercise 16, because $f(z)$ is included in a path of $\frac{1}{(z-1)^2(\frac{z-1}{2})}$, it is calculated as

$$\int_c f(z)dz = 2\pi i\left(\underset{z=a}{Res} f(z) + \underset{z=\frac{1}{2}}{Res} f(z)\right).$$

5. Application of real number integration of Residue theorem

There are many numbers that can be easily integrated with Residue theorem. For example:

1) Integration of rational function of trigonometrical function

Think about the integration of the following formula:

$$I=\int_0^{2\pi} R(\cos\theta,\sin\theta)d\theta \tag{6-82}$$

Provided that, function R is a rational function of real number in $0 \leq \theta \leq 2\pi$. If $e^{i\theta}=z$,

$$\cos\theta=\frac{1}{2}(e^{i\theta}+e^{-i\theta})=\frac{1}{2}\left(z+\frac{1}{z}\right)$$

$$\sin\theta=\frac{1}{2i}(e^{i\theta}-e^{-i\theta})=\frac{1}{2}i\left(z-\frac{1}{z}\right) \tag{6-83}$$

If θ changes from 0 to 2π, $z=e^{i\theta}$ turns around the unit circle of $|z|=1$ counter-clockwise.

$$\frac{dz}{d\theta}=ie^{i\theta},\ d\theta=\frac{dz}{iz} \tag{6-84}$$

If the formula (6-82) is written as the following integral equation on a complex plane:

$$I=\int_c f(z)\cdot\frac{dz}{iz} \tag{6-85}$$

Where, the integrand $f(z)/iz$ must not have a pole on the unit circumference.

Exercise 6-16

Evaluate $I=\int_c^{2\pi}\frac{d\theta}{1-2p\cos\theta+p^2}(0<p<1)$.

Solution

Using (6-83), it can be converted into the following complex function:

$$I=\int_c^{2\pi}\frac{dz/iz}{1-2p(z+\frac{1}{z})+p^2}=\int_c\frac{dz}{i(1-pz)(2-p)}$$

Where, the poles of the integrand are $\frac{1}{p}(>1)$ and $P(<1)$. The pole included in the unit circle c is $z=P$ only. Therefore the residue is as follows:

$$\underset{z=P}{Res}\frac{1}{i(1-pz)(z-p)}=\lim_{z=0}(z-p)\frac{1}{i(1-pz)(z-p)}$$
$$=\frac{1}{i(1-p^2)}$$

And according to Residue theorem, the result is as follows:

$$I=2\pi i\underset{z=p}{Res}f(z)=\frac{2\pi}{1-p^2}$$

2) Improper integral of rational function

Integration with the infinite integration section is called an improper integral, and is represented as follows:

$$\int_{-\infty}^{+\infty}f(x)dx=\lim_{r\to\infty}\int_{-r}^{r}f(x)dx \tag{6-86}$$

If the degree is higher than the denominator of $f(x)$ by at least 2 degrees, and the denominator has no real root, it can be the integration $\int_c f(z)dz$ of the upper half of the circle with the radius r of Figure (6-16).

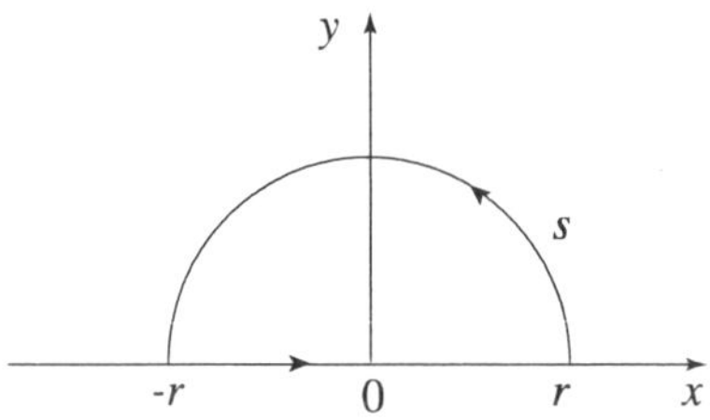

Figure 6-15 Improper integral

Where r is extended sufficiently to include all the poles in the upper half.

According to Residue theorem, the function is as follows:

$$\int_c f(z)dx = \int_s f(z) + \int_{-r}^{r} f(x)dx$$

$$= 2\pi\sum Resf(z) \tag{6-87}$$

Therefore, the real number integration is as follows:

$$\int_{-r}^{r} f(x)dx = 2\pi i\sum Resf(z) - \int_s f(z)dz \tag{6-88}$$

In the integral value following the semicircumference s in Formula (6-88), the denominator of $f(z)$ is higher than the numerator by at least 2 degrees. Therefore, the value is 0 if $r\to\infty$. And Formula (6-88) can be writtec as follows:

$$\int_{\infty}^{-\infty} f(x)dx = 2\pi\sum Resf(z) \tag{6-89}$$

Exercise 6-17

Prove that $\int_{\infty}^{-\infty}\frac{x^3dx}{(x^2+1)(x^2+2x+2)} = \frac{7\pi}{50}$.

Solution

$f(z) = \frac{z^2}{(z^2+1)(z^2+2z+2)}$ has a secondary pole at $z=i$, and a primary pole at $z=(-1+i)$ and $(-1-i)$. In the following diagram, $z=-1-i$ is excluded as it is not included in the semicircumference of the upper half. Therefore,

$$\int_{\infty}^{-\infty} f(x)dx = 2\pi i\left\{\underset{z=i}{Res} f(z) + \underset{z=-i+i}{Res} f(z)\right\}$$

$$\underset{z=i}{Res} f(z) = \lim_{z\to i}\frac{d}{dz}\left\{(z-i)^2\frac{z^2}{(z+i)^2(z-i)^2(z^2+2z+2)}\right\}\frac{9i-12}{100}$$

$$\underset{z=-i+i}{Res} f(z) = \lim_{z=-1+i} (z+1-i) \cdot \frac{z^2}{(z^2+1)^2(z+1-i)(z+1+i))} = \frac{3-4i}{25}$$

$$\therefore \int_{-\infty}^{\infty} \frac{x^2 dx}{(x^2+1)^2(x^2+2x+2)} = 2\pi i\left(\frac{9i-12}{100} + \frac{3-4i}{25}\right) = \frac{7\pi}{50}.$$

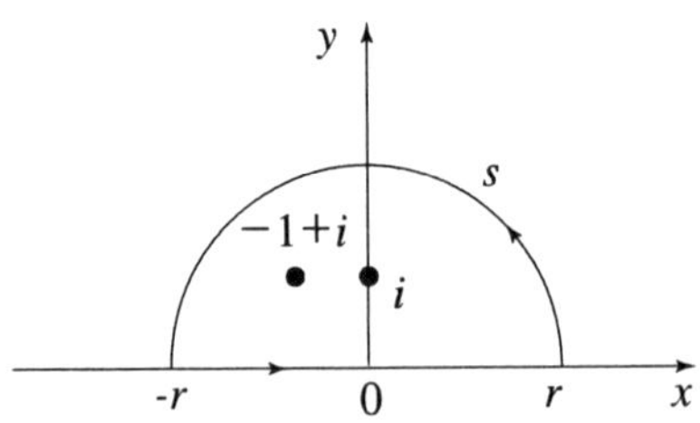

Chapter 7

Laplace Transform

1. Definition of Laplace transform

For a function $f(t)$ defined at $t > 0$, the following integral transform with s as the variable is valid:

$$F(s) = \mathcal{L}[f(t)] = \int_0^\infty f(t)e^{-st}dt \tag{7-1}$$

If there is an integral value, $F(s)$ is called the Laplace transform of a given function $f(t)$. The following formula used to evaluate $f(t)$ with given $F(s)$ is called inverse Laplace transform:

$$f(t) = \frac{1}{2\pi j}\int_{c-j\infty}^{c+j\infty} F(s)e^{st}ds \tag{7-2}$$

It is written as $\mathcal{L}^{-1}[F(s)]$, that is

$$f(t) = \mathcal{L}^{-1}[F(s)] \tag{7-3}$$

2. Laplace transform of important function

1) Unit step function: $u(t)$

$$u(t) = \begin{cases} 1, t \geq 0 \\ 0, t < 0 \end{cases}$$

$$\mathcal{L}[u(t)] = \int_0^\infty u(t)e^{-st}dt = \int_0^\infty e^{-st} = -\frac{1}{s}e^{-st}\Big|_0^\infty$$

$$= \frac{1}{s}\,(s > 0) \tag{7-4}$$

2) $e^{at}u(t)$

$$\mathcal{L}[(e^{at})u(t)] = \int_0^\infty e^{-s}e^{at}dt = \frac{1}{a-s}e^{-(s-a)t}\Big|_0^\infty$$

$$= \frac{1}{s-a}(s-a>0)\ \text{(7-5)}$$

3) Sine function: $\sin\omega t$

Because $\sin\omega t = \frac{1}{2}j(e^{j\omega t} - e^{-j\omega t})$,

$$\mathcal{L}[\sin\omega t] = \int_0^\infty \frac{1}{2j}(e^{j\omega t} - e^{-j\omega t})\cdot e^{-st}dt$$

$$= \frac{1}{2j}\left(\frac{1}{s-j\omega} - \frac{1}{s+j\omega}\right)$$

$$= \frac{1}{2j}\left(\frac{2j\omega}{s^2+\omega^2}\right) = \frac{\omega}{s^2+\omega^2} \tag{7-6}$$

4) Cosine function: $\cos\omega t$

$$\cos\omega t = \frac{1}{2}(e^{j\omega t} + e^{-j\omega t})$$

Therefore,

$$\mathcal{L}[\cos\omega t] = \int_0^\infty \frac{1}{2}(e^{j\omega t} + e^{j-\omega t})\cdot e^{-st}dt$$

$$= \frac{1}{2}\left(\frac{1}{s-j\omega} + \frac{1}{s+j\omega}\right)$$

According to (7-5),

$$= \frac{1}{2}\frac{2s}{s^2+\omega^2} = \frac{s}{s^2+\omega^2} \qquad (7\text{-}7)$$

5) $\sinh at$ and $\cosh at$

$$\sinh at = \frac{1}{2}(e^{at} - e^{-at})$$

$$\cosh at = \frac{1}{2}(e^{at} + e^{-at}),$$ therefore

$$\mathcal{L}[\sinh at] = \frac{a}{s^2-a^2} \qquad (7\text{-}8)$$

$$\mathcal{L}[\cosh at] = \frac{s}{s^2-a^2} \qquad (7\text{-}9)$$

6) $e^{at}\sin\omega t$

Theorem

Laplace transform of a function multiplied by e^{at} equals to the Laplace transform formula where s is replaced by $s-a$. This is called a complex shifting theorem. For example, $\mathcal{L}[f(t)] = F(s)$ is transformed to the following:

$$\mathcal{L}[e^{at}f(t)] = F(s-a) \qquad (7\text{-}10)$$

Proof

$$\mathcal{L}[e^{at}f(t)] = \int_0^\infty e^{-st}[e^{at}f(t)]dt$$

$$= \int_0^\infty e^{-(s-a)t}f(t)dt$$

$$= F(s-a)$$

Using (7-6) and (7-10),

$$\mathcal{L}[e^{at}\sin\omega t] = \frac{\omega}{(s-a)^2+\omega^2} \tag{7-11}$$

7) $e^{at}\cos\omega t$

Using (7-7) and (7-10),

$$\mathcal{L}[e^{at}\cos\omega t] = \frac{s-a}{(s-a)^2+\omega^2} \tag{7-12}$$

Table 7-1 Laplace transform of elementary functions

$f(t)$	$F(s) = \mathcal{L}[f(t)] \equiv \int_0^\infty f(t)e^{st}dt$
(1) $u(t)$	$\frac{1}{s}$
(2) e^{at}	$\frac{1}{s-a}$
(3) $\sin\omega t$	$\frac{\omega}{s^2+\omega^2}$
(4) $\cos\omega t$	$\frac{s}{s^2+\omega^2}$
(5) $\sinh at$	$\frac{a}{s^2-a^2}$
(6) $\cosh at$	$\frac{s}{s^2-a^2}$
(7) $e^{at}f(t)$	$F(s-a)$

Theorem

When $\mathcal{L}[f(t)] = F(s)$,

$$\mathcal{L}[t^n f(t)] = (-1)^n F^{(n)}(s)\ (n=1,2,\cdots\cdots) \tag{7-13}$$

Proof

$$F(s)=\mathcal{L}[f(s)]=\int_0^\infty F(t)e^{-st}dt$$

Differentiate both sides for s to prove as follows:

$$F'(s)=\frac{d}{ds}\int_0^\infty f(t)e^{-st}dt=\int_0^\infty \frac{\partial}{\partial s}[e^{-st}f(t)]dt$$

$$=\int_0^\infty(-te^{-st})f(t)dt=\int_0^\infty[-tf(t)]e^{-st}dt$$

$$=\mathcal{L}[-tf(t)]=-\mathcal{L}[tf(t)]$$

$$F''(s)=\int_0^\infty\frac{\partial}{\partial s}[\{-tf(t)\}e^{-st}]dt$$

$$=(-1)^2\int_0^\infty[t^2f(t)]e^{-st}dt$$

$$=(-1)^2\mathcal{L}[t^2f(t)]$$

$$\vdots$$

$$F^n(s)=(-1)^n\mathcal{L}[t^nf(t)]$$

Exercise 7-1

Evaluate the Laplace transforms.

(a) $t\sin at$ (b) $t^2\sin 2t$ (c) $t^2\cos at$

Solution

(a) Because $\mathcal{L}[\sin at]=\frac{a}{s^2+a^2}$, using the theorem,

$$\mathcal{L}[t\sin at]=-\frac{d}{ds}\frac{a}{s^2+a^2}=\frac{2as}{(s^2+a^2)^2}$$

(b) Because $\mathcal{L}[\sin 2t]=\frac{2}{s^2+2^2}$, using the theorem,

$$\mathcal{L}\,[t^2\sin 2t] = (-1)^2\frac{d^2}{ds^2}\left(\frac{2}{s^2+4}\right) = \frac{10s^2-16}{(s^2+4)^3}$$

(c) Because $\mathcal{L}\,[\cos at] = \dfrac{s}{s^2+a^2}$, using the theorem,

$$\mathcal{L}\,[t^2\cos at] = (-1)^2\frac{d^2}{ds^2}\left(\frac{s}{s^2+a^2}\right) = \frac{2s(3a^2-s^2)}{(s^2+a^2)}$$

Theorem

When $\mathcal{L}\,[f(t)] = F(s)$,

$$\mathcal{L}\left[\frac{f(t)}{t}\right] = \int_s^\infty F(s)ds \qquad (7\text{-}14)$$

Proof

Integrate both sides of $F(s) = \mathcal{L}\,[f(t)] = \int_0^\infty f(t)e^{-st}dt$ from s to ∞. On the assumption that it is possible to change the order of integration of multiple integral,

$$\int_s^\infty F(s)ds = \int_0^\infty\left[\int_s^\infty f(t)e^{-st}ds\right]dt = \int_0^\infty f(t)\left[\int_s^\infty e^{-st}ds\right]dt$$

$$= \int_0^\infty f(t)\left[\frac{e^{-st}}{-t}\right]_s^\infty dt = \int_0^\infty \frac{f(t)}{t}e^{-st}dt = \mathcal{L}\left[\frac{f(t)}{t}\right]$$

Therefore, this theorem may be extended to general cases through continued integration of $F(s)$.

Exercise 7-2

Evaluate $\mathcal{L}\left[\frac{\sin at}{t}\right]$.

Solution

$$\mathcal{L}[\sin at]=\frac{a}{s^2+a^2}$$

Therefore, using the theorem,

$$\mathcal{L}\left[\frac{\sin at}{t}\right]=\int_s^\infty \frac{a}{s^2+a^2}ds=\left[\tan^{-1}\frac{s}{a}\right]_s^\infty$$

$$=\frac{\pi}{2}-\tan^{-1}\frac{s}{a}=\cos t^{-1}\frac{s}{a}$$

Exercise 7-3

Evaluate inverse Laplace transform of function $F(s)=\frac{s}{(s^2-1)^2}$.

Solution

$$\mathcal{L}[f(t)]=\frac{s}{(s^2-1)^2}$$

$$\mathcal{L}\left[\frac{f(t)}{t}\right]=\int_s^\infty \frac{s}{(s^2-1)^2}ds=\frac{1}{2(s^2-1)}\Big|_s^\infty$$

$$=\frac{1}{4}\left(\frac{1}{s-1}-\frac{1}{s+1}\right)\Big|_s^\infty$$

Therefore,

$$\frac{f(t)}{t}=\mathcal{L}^{-1}\left[\frac{1}{4}\left(\frac{1}{s-1}-\frac{1}{s+1}\right)\right]$$

$$=\frac{e^t-e^{-t}}{4}=\frac{\sinh t}{2}$$

And

$$f(t)=\frac{t\sinh t}{2}.$$

---

3. Basic theorems of Laplace transform

Linearity

Laplace transform is a linear operation. If there is a Laplace transform for function $f(t)$ and $g(t)$, respectively, and a and b are constants, the following formula is valid.

$$\mathcal{L}\,[af(t)+bg(t)]=aF(s)+bG(s) \tag{7-15}$$

Where, $F(s)$ and $G(s)$ are Laplace transform functions of $f(t)$ and $g(t)$, respectively.

Laplace transform of derivatives: Laplace transform of the 1st derivation of a time function $f(t)$ equals to $f(\alpha)$ Laplace transform multiplied by s minus the limit of $f(t)$ when t approaches to 0. In other words,

$$\mathcal{L}\left[\frac{df(t)}{dt}\right]=sF(s)-\lim_{t\to 0}f(t)=sF(s)-f(0) \tag{7-16}$$

Thinking about higher-order differential equations,

$$\begin{aligned}\mathcal{L}\left[\frac{d^n f(t)}{dt^n}\right]&=s^nF(s)-\lim_{t\to 0}\left[s^{n-1}f(t)+s^{n-2}\frac{dt(t)}{dt}+\cdots\cdots+\frac{d^{n-1}f(t)}{dn^{n-1}}\right]\\&=s^nF(s)-s^{n-1}f(0)-s^{n-2}f^{(1)}(0)-\cdots\cdots-f^{(n-1)}(0)\end{aligned} \tag{7-17}$$

Laplace transform of integral function

Laplace transform of the 1st integral of a time function $f(t)$ equals transform of $f(t)$ divided by s.

$$\mathcal{L}\left[\int_0^t f(\tau)d\tau\right] = \frac{F(s)}{s} \tag{7-18}$$

In general, nth integration is as follows:

$$\mathcal{L}\left[\int_0^{t_1}\int_0^{t_2}\cdots\cdots\int_0^{t_n} f(\tau)d\tau dt_1 \cdots\cdots dt_{m-1}\right] = \frac{F(s)}{s^n} \tag{7-19}$$

Time shifting

Laplace transform of function $f(t)$ delayed by time T equals Laplace transform of $f(t)$ multiplied by e^{-TS}.

$$\mathcal{L}[f(t-T)u_s(t-T)] = e^{-Ts}F(s) \tag{7-20}$$

Where $u_s(t-T)$ is a unit step function shifted right by time T.

Theorem of initial value

If the Laplace transform function of $f(t)$ is $F(s)$, and there is a following time limit, the initial value is as follows:

$$\lim_{t\to 0} f(t) = \lim_{s\to\infty} sF(s) \tag{7-21}$$

Theorem of final value

If the Laplace transform function of $f(t)$ is $F(s)$, and is analytic on an imaginary axis and within right-half-plane on plane s, the final value is as follows:

$$\lim_{t\to\infty} f(t) = \lim_{s\to 0} sF(s) \tag{7-22}$$

According to the theorem of final value, it is possible to evaluate the final value of time function by using Laplace transform at $s \to 0$. In thus case, it is useful to analyze the feedback control system. However, this theorem is not valid as it is not analytic if the real part of $sF(s)$ has a pole of 0 or a positive number.

Complex shifting

The Laplace transform of the function $f(t)$ multiplied by $e^{\mp at}$ equals to the Laplace transform function $F(s)$ in which $s \pm a$ is substituted for s:

$$\mathcal{L}[e^{\mp at} f(t)] = F(s \pm a) \tag{7-23}$$

Where a is a constant.

Table 7-2 Theorems on Laplace transform

Multiplication by a constant	$\mathcal{L}[kf(t)] = kF(s)$
Addition and subtraction	$\mathcal{L}[f_1(t) \pm f_2(t)] = F_1(s) \pm F_2(s)$ $\mathcal{L}\left[\frac{df(t)}{dt}\right] = sF(s) - f(0)$
Differential	$\mathcal{L}\left[\frac{d^n f(t)}{dt^n}\right] = s^n F(s) - s^{n-1} f(0) - s^{n-2} f^{(1)}(0)$ $- \cdots\cdots - s f^{(n-2)}(0) - f^{(n-1)}(0)$ where, $f(k)(0) = \left.\frac{d^k f(t)}{dk^k}\right\vert_{t=0}$

Table 7-2 continued

Integral	$\mathcal{L}\left[\int_0^t f(\tau)d\tau\right] = \frac{F(s)}{s}$ $\mathcal{L}\left[\int_0^{t_1}\int_0^{t_2}\int_0^{t_3} f(\tau)d\tau dt_1 \cdots\cdots dt_{n-1}\right] = \frac{F(s)}{s^n}$
Time shifting	$\mathcal{L}[f(t-T)u_s(t-T)] = e^{-Ts}F(s)$
Theorem of initial value	$\lim_{t\to\infty} f(t) = \lim_{s\to\infty} sF(s)$
Multiplication by constant	$\mathcal{L}[kf(t) = kF(s)$
Theorem of final value	$\lim_{t\to\infty} f(t) = \lim_{s\to 0} sF(s)$ if $sF(s)$ has no poles on the right of or above the axis.
Complex shifting	$\mathcal{L}[e^{\mp at}f(t)] = F(s \pm a)$
Convolution integral	$F_1(s)F_2(s) = \mathcal{L}\left[\int_0^t f_1(\tau)f_2(t-\tau)d\tau\right]$ $= \mathcal{L}\left[\int_0^t f_2(\tau)f_1(t-\tau)d\tau\right]$ $= \mathcal{L}[f_1(t)*f_2(t)]$

4. Laplace transform of differential and integral function

The following theorem of Laplace transform of differential and integral of the function $f(t)$ plays an important role in solving differential equations, and in evaluating Laplace transform and inverse Laplace transform.

Theorem

If $f(t)$ and its derivatives are continuous or partially continuous at $t > 0$, and

$$\lim_{t\to\infty} f(t)e^{-st} = 0, \lim_{t\to\infty} f(t) = f(0^+),$$

$$\mathcal{L}\left[\frac{df(t)}{dt}\right] = \mathcal{L}\,[f(t)] = \mathcal{L}\,[f(t)] - f(0^+)$$

$$= sF(s) - f(0^+) \tag{7-24}$$

Proof

From the definition,

$$F(s) = \mathcal{L}\,[f(t)] = \int_0^\infty f(t)e^{-st}dt$$

Therefore, using the following:

$$u = f(t), du = [df(t)]$$

$$dv = e^{-st}dt, v = -\frac{1}{s}e^{-st}$$

$F(s)$ is evaluated as the following:

$$F(s) = -\frac{1}{s}f(t)e^{-st}\Big|_0^\infty + \frac{1}{s}\int_0^\infty \left[\frac{df(t)}{dt}\right]e^{-st}dt$$

$$= \frac{f(0^+)}{s} + \frac{1}{s}\mathcal{L}\left[\frac{df(t)}{dt}\right]$$

Therefore,

$$\mathcal{L}\left[\frac{df(t)}{dt}\right] = sF(s) - f(0^+)$$

and

$$\mathcal{L}\,[f''(t)] = s\mathcal{L}\,[f'(t)] - f'(0^+)$$

$$= s[sF(s) - f(0^+)] - f'(0^+)$$

$$= s^2F(s) - sf(0^+) - f'(0^+) \tag{7-25}$$

In general,

$$\mathcal{L}\,[f^n(t)] = \mathcal{L}\left[\frac{d^n f(t)}{dt}\right]$$

$$= s^nF(s) - s^{n-1}f(0^+) - s^{n-2}f'(0^+)\cdots$$

$$- sf^{(n-2)}(o^+) - f^{(n-1)}(0^+)$$

$$= s^n F(s) - \sum_{k=1=0}^{n} s^{n-k} f^{(k-1)}(0^+) \quad (7\text{-}26)$$

Exercise 7-4

If $f(t)=t^2$, find the Laplace transform of the function.

$$f(0^+) = f'(0^+) = 0$$

$$f''(t) = 2$$

Solution

$$\mathcal{L}[2] = 2/s$$

From the theorem,

$$\mathcal{L}[f'(t)] = \mathcal{L}[2]$$

$$s^2\mathcal{L}[t^2] - sf(0^+) - f(0^+) = \frac{2}{s}.$$

Therefore,

$$\mathcal{L}[t^2] = \frac{2}{s^3}.$$

Exercise 7-5

Evaluate $\mathcal{L}[f(t)]$ if $f(t) = \sin^2 t$.

Solution

$$f'(t) = 2\sin t\cos t = \sin 2t$$

Apply Laplace transform for both sides.

$$\mathcal{L}[f'(t)] = \mathcal{L}[\sin 2t]$$

$$sF(s) - f(0^+) = \frac{2}{s^2+4} f(0^+) = 0$$

Therefore,

$$F(s) = \mathcal{L}[\sin^2 t] = \frac{2}{s(s^2+4)}$$

Exercise 7-6

Evaluate the Laplace transform of $f(t)$ when $f(t) = t\sin t$.

Solution

$$f'(t) = \sin t + t\cos t \quad f(0^+) = f'(0^+) = 0$$

$$f''(t) = 2\cos t - t\sin t = 2\cos t - f(t)$$

If both sides are Laplace-transformed,

$$\mathcal{L}[f''(t)] = 2\mathcal{L}[\cos t] - \mathcal{L}[f(t)]$$

$$s^2F(s) - sf(0^+) - f'(0^+) = 2\frac{s}{s^2+1} - F(s).$$

Therefore,

$$F(s) = \frac{2s}{(s^2+1^2)}$$

Theorem

If $f(t)$ is continuous or partially continuous, and

$$\lim_{t\to\infty} e^{-st}\int_0^\infty f(t)dt = 0,$$

then,

$$\mathcal{L}\left[\int_0^t f(\tau)d\tau\right] = \frac{\mathcal{L}[f(t)]}{s} = \frac{F(s)}{s} \tag{7-27}$$

Proof

According to the theorem,

$$\mathcal{L}\left[\int_0^t f(\tau)d\tau\right] = \int_0^\infty \left[\int_0^t f(\tau)d\tau\right] e^{-st}dt$$

If a partial integration is applied,

$$u = \int_0^t f(\tau)d\tau, \quad du = f(t)dt$$

$$dv = e^{-st}dt \quad v = -\frac{1}{s}e^{-st}$$

$$\mathcal{L}\left[\int_0^t f(\tau)d\tau\right] = -\frac{e^{-st}}{s}\int_0^t f(\tau)d\tau \Big|_s^\infty + \frac{1}{s}\int_s^\infty f(t)e^{-st}dt$$

$$= \frac{1}{s}\mathcal{L}[f(t)] = \frac{F(s)}{s}$$

Using (7-27), the Laplace transform of the indefinite integral is represented as the following:

$$\int f(t)dt = \int_0^t f(\tau)d\tau + \lim_{t\to 0+} f(t)$$

$$= \int_0^t f(z)dz + f(0^+)$$

If Laplace transform is applied to the indefinite integral,

$$\mathcal{L}\left[\int f(t)dt\right]=\mathcal{L}\left[\int_0^t f(\tau)d\tau\right]+\mathcal{L}[f(0^-)]$$

$$=\frac{F(s)}{s}+\frac{f(0^-)}{s} \tag{7-28}$$

Exercise 7-7

Evaluate $f(t)$ of the following functions. In other words, evaluate the inverse Laplace transform.

(a) $\mathcal{L}[f(t)]=\dfrac{a}{s(s+a)}$ (b) $\mathcal{L}[f(t)]=\dfrac{1}{s^2(s^2+\omega^2)}$

Solution

(a) $\mathcal{L}[e^{-at}]=\dfrac{1}{s+a}$

Therefore, from the theorem,

$$\mathcal{L}\left[\int_0^t e^{-st}dt\right]=\frac{1}{s(s+a)}$$

$$\mathcal{L}\left[\int_0^t e^{-st}dt\right]=\mathcal{L}\left[-\frac{1}{a}e^{-at}rightvert_0^t\right]=\frac{1}{a}\mathcal{L}[1-e^{-st}]$$

Therefore,

$$\mathcal{L}[1-e^{-at}]=\frac{a}{s(s+a)}$$

or

$$f(t)=1-e^{-at}.$$

(b) $\mathcal{L}^{-1}\left[\dfrac{1}{(s^2+\omega^2)}\right]=\dfrac{1}{\omega}sin\omega t$

Therefore,

$$\mathcal{L}^{-1}\left[\frac{1}{(s(s^2+\omega^2)}\right]=\frac{1}{\omega}\int_0^t \sin\omega\tau d\tau=\frac{1}{\omega^2}(1-\cos\omega t)$$

From the theorem,

$$f(t) = \mathcal{L}^{-1}\left[\frac{1}{s^2(s^2+\omega^2)}\right] = \frac{1}{\omega^2}\int_0^t (1-\cos\omega\tau)\,d\tau$$

$$= \frac{1}{\omega^2}\left[t - \frac{\sin\omega t}{\omega}\right].$$

5. Laplace transform of signal function

1) Unit step function

As shown in Figure (7-1), the following function is called a unit step function and is denoted by $u(t)$:

$$u(t) = \begin{cases} 1, & t \geq 0 \\ 0, & t > 0 \end{cases} \tag{7-29}$$

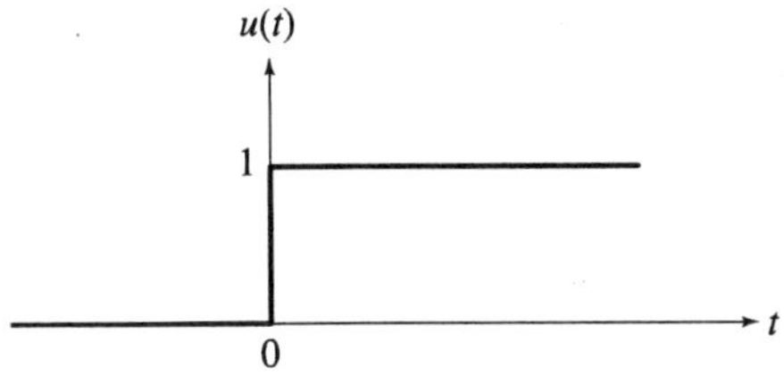

Figure 7-1 Unit step function

Formula (7-29) is a function, as shown in Figure (7-1), that changes suddenly from 0 to the unit value at $t=0$. Figure (7-2) shows the generalized unit step function. This function is represented as follows:

$$u(t-a)=\begin{cases}1, & t \geq a \\ 0, & t > a\end{cases} \tag{7-29}$$

Formula (7-30) is a step function that changes instantly at $t=a$ as shown in Figure (7-2). In general, a step function has a unit value for positive independent variable $(t-a)$ of function u, and 0 for negative $(t-a)$.

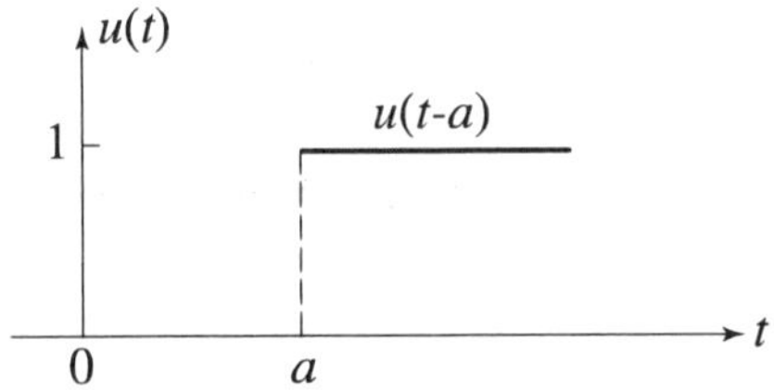

Figure 7-2 Generalized unit step function

The Laplace transform of $f(t)=u(t)=1$ is as the follows:

$$F(s)=\int_0^\infty u(t)e^{-st}dt=\int_0^\infty 1\cdot e^{-st}dt=-\frac{1}{s}\int_0^\infty(-s)e^{-st}dt$$

$$=-\frac{1}{s}[e^{-st}]_0^\infty=\frac{1}{s} \tag{7-31}$$

For Laplace transform of $f(t)=u(t-a) is, if t-a=\tau$,

$$F(s)=\int_0^\infty u(t-a)e^{-st}dt=\int_0^\infty u(\tau)e^{-as}d\tau=\frac{1}{s}e^{-as} \tag{7-32}$$

2) Shifting unit step function

As in Figure(7-3)(a), $u(t+a)$ is a function which changes by unit value at the moment of $t=-a$. As in Figure (7-3)(b), $u(t-a)$ is a function which changes from the unit value to 0 at the moment of $t=a$.

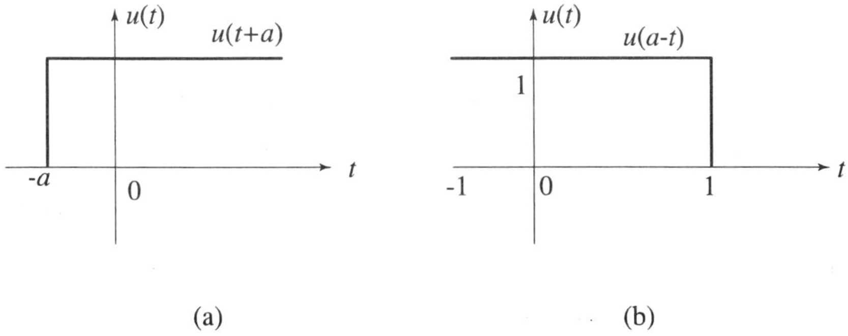

Figure 7-3 Shifting unit step function

Unit step function is applicable to a time function $f(t)$. A function with the time delay of $t=a$, that is a shifting time function, is represented as the following:

$$f(t-a)u(t-a) \tag{7-33}$$

To evaluate the Laplace transform of this function, introduce a new variable t'.

$$F(s)=\int_0^{\infty} f(t')e^{-st}dt \tag{7-34}$$

If $t'=t-a$, formula (7-34) is as follows:

$$F(s)=\int_0^{\infty} f(t-a)e^{-(t-a)s}dt=\int_a^{\infty} f(t-a)e^{-st}\cdot e^{as}dt$$

If $f(t-a)$ is multiplied by $u(t-a)$, the minimum of the integral can be changed to 0.

$$F(s)=e^{as}\int_0^{\infty} f(t-a)u(t-a)e^{-st}dt \tag{7-35}$$

Because the integral equation is a transform of function $f(t-a)u(t-a)$, (7-35) is represented as follows:

$$e^{-as}\mathcal{L}[f(t)] = \mathcal{L}[f(t-a)u(t-a)] \qquad (7\text{-}36)$$

Therefore, the inverse Laplace transform formula is as follows:

$\mathcal{L}^{-1}[e^{as}F(s)] = f(t-a)u(t-a)$ (7-36)

From this formula, we see that the transformation of the function delayed by $t=a$ equals the Laplace transform of the function starting at $t=0$ multiplied by e^{-as}. This is called a shifting theorem.

3) Unit ramp function

A unit ramp function is a straight line function with a slope of 1 as shown in Figure (3-4).

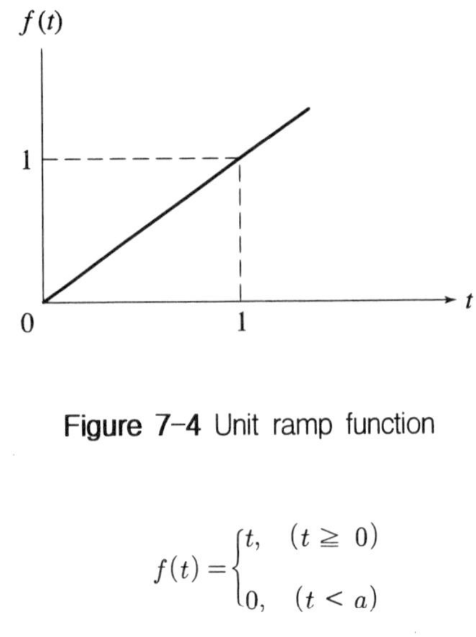

Figure 7-4 Unit ramp function

$$f(t) = \begin{cases} t, & (t \geq 0) \\ 0, & (t < a) \end{cases} \qquad (7\text{-}38)$$

The Laplace transform is as follows:

$$\mathcal{L}[f(t)] = \int_0^{\infty} tu(t)e^{-st}dt = \int_0^{\infty} te^{-st}dt \qquad (7\text{-}39)$$

If $u=t, v'=e^{-st}$ for partial integration,

$$u' = 1, v = -\frac{1}{s}e^{-st}$$

If they are substituted in the partial integral equation,

$$\mathcal{L}[f(t)] = \left[t \cdot \frac{e^{-st}}{-s}\right]_0^\infty - \int_0^\infty \frac{e^{-st}}{-s} dt$$

$$= \frac{1}{s}\int_0^\infty e^{-st} dt = -\frac{1}{s^2}[e^{-st}]_0^\infty$$

$$= \frac{1}{s^2}$$

Therefore,

$$F(s) = \frac{1}{s^2} \tag{7-40}$$

4) Unit impulse function

A limit waveform of $\epsilon \to 0$ for a wave form with the width of ϵ, height of $\frac{1}{\epsilon}$ and an area of 1, as shown in Figure (7-5), is a unit impulse function, and is represented as $\delta(t)$.

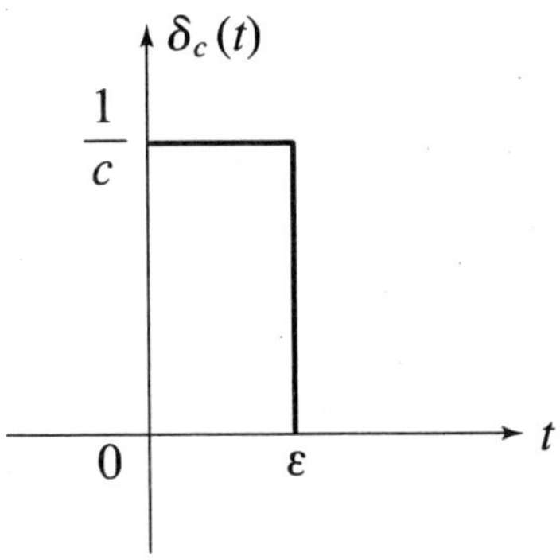

Figure 7-5 Unit impulse function

The Laplace transform of a unit impulse function $\delta(t)$ is represented as follows, using a unit step function:

$$\delta_\epsilon = \frac{1}{\epsilon}\{u(t) - u(t-\epsilon)\} \tag{7-41}$$

According to the definition,

$$\delta(t) = \lim_{\epsilon \to \infty}\delta_\epsilon(t) = \frac{du(t)}{dt} \tag{7-42}$$

In other words, a unit impulse function $\delta(t)$ is a unit step function $u(t)$ integrated with time t.

Therefore,

$$\mathcal{L}[\delta(t)] = \mathcal{L}\lim_{\epsilon \to \infty}\delta_\epsilon(t) = \lim_{\epsilon \to 0}\int_0^\infty \delta_\epsilon(t)e^{-st}dt$$

$$= \lim_{\epsilon \to 0}\frac{1}{\epsilon}\int_0^\infty \{u(t) - u(t-\epsilon)e^{-st}dt\}dt$$

$$= \lim_{\epsilon \to 0}\frac{1}{\epsilon}\left(\frac{1-e^{-\epsilon s}}{s}\right) = 1 \quad (7\text{-}43)$$

6. Laplace transform of periodic function

The time shifting theorem is an important theorem that is applicable to the all the engineering fields.

Theorem

If $\mathcal{L}[f(t)] = F(s)$, for every positive number a,

$$\mathcal{L}[f(t-a)u_{a(t)}] = e^{-as}F(s) \tag{7-44}$$

This is called a time shifting theorem.

Proof

$$\mathcal{L}[f(t-a)u_a(t)] = \int_0^{\infty} [f(t-a)u_a(t)e^{-st}]dt$$

$$= \int_0^{\infty} f(t-a)e^{-at}dt$$

$$f(t-a)u_{a(t)} = \begin{cases} 0 & t < a \\ f(t-a) & t \geqq 0 \end{cases}$$

Therefore,

$t-a=\tau \quad dt=d\tau$

$\tau=0$ if $t=a$

Therefore,

$$\mathcal{L}[f(t-a)u_a(t)] = \int_0^{\infty} f(\tau)e^{-s(\tau+a)}d\tau$$

$$= e^{-sa}\int_0^{\infty} f(\tau)e^{-s\tau}d\tau$$

$$= e^{-sa}F(s)$$

Check out the following 4 waveforms:

(1) $\sin\omega(t-a)$

(2) $\sin\omega(t-a)u_0(t)$

(3) $\sin\omega t u_a(t)$

(4) $\sin\omega(t-a)u_a(t)$

As shown in Figure (7-6), they make different waveforms.

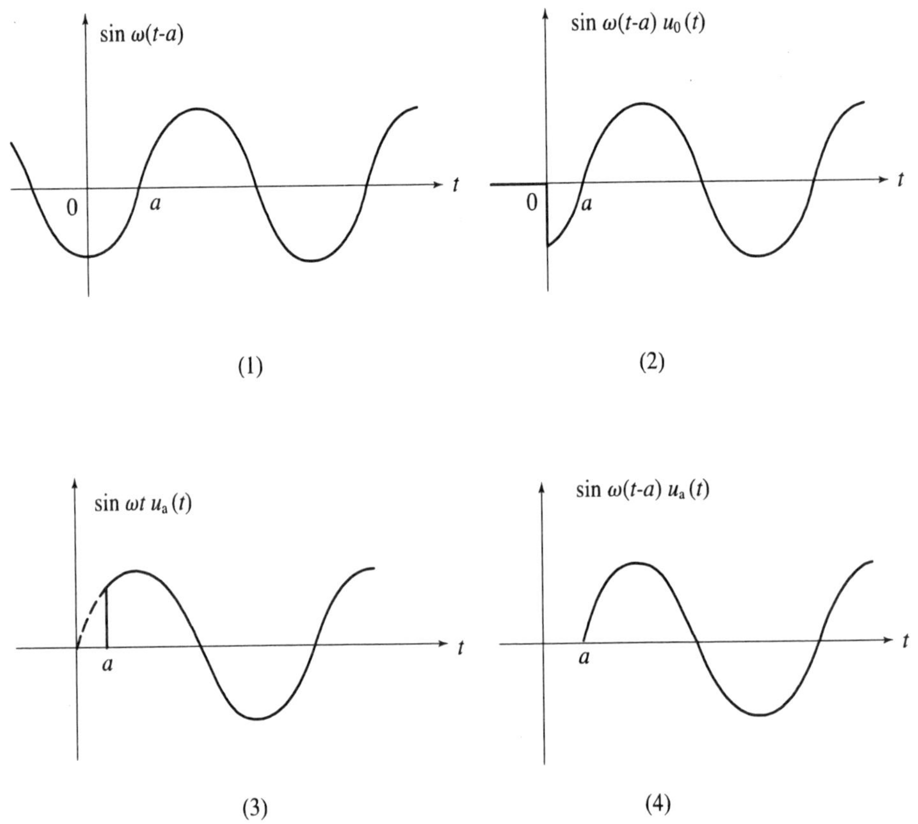

Figure 7-6 Shifted sine waves

Laplace transforms of these 4 waveforms are evaluated as follows: Because the integration limit of a Laplace transform is from 0 to ∞ as described in the definition of a Laplace transform in the previous section, the results of Laplace transform for (1) and (2) are as follows:

$$\mathcal{L}\,[\sin\omega(t-a)] = \mathcal{L}\,[\sin\omega t\cos\omega a - \cos\omega t\sin\omega a]$$

$$= \frac{\omega\cos\omega a - s\sin\omega a}{s^2} \qquad (7\text{-}45)$$

For the waveform of (3), the following is made:

$$\mathcal{L}[\sin\omega t u_a(t)] = \int_0^\infty \sin\omega u_a(t)e^{-st}dt$$

$$= \int_a^\infty \sin\omega t e^{-st}dt$$

$$= \frac{1}{2j}\int_a^\infty [e^{(-s+j\omega)t)}e^{(-s-j\omega)t}dt$$

$$= e^{-as}\frac{\omega\cos\omega a + s\sin\omega a}{s^2+\omega^2} \quad (7\text{-}46)$$

For the waveform of (4), the following is made according to the shifting theorem:

$$\mathcal{L}[\sin\omega(t-a)u_a(t)] = e^{-as}\mathcal{L}[\sin\omega t]$$

$$= e^{-as}\frac{\omega}{s^2+\omega^2} \quad (7\text{-}47)$$

The above results show that the Laplace transforms of these waveforms are different from each other. Using the time shifting theorem and Table (7-1), we can easily evaluate the Laplace transforms, as described in the following example, without integrating the actual waveform.

Exercise 7-8

Evaluate Laplace transform of the sawtooth wave in Figure (7-7).

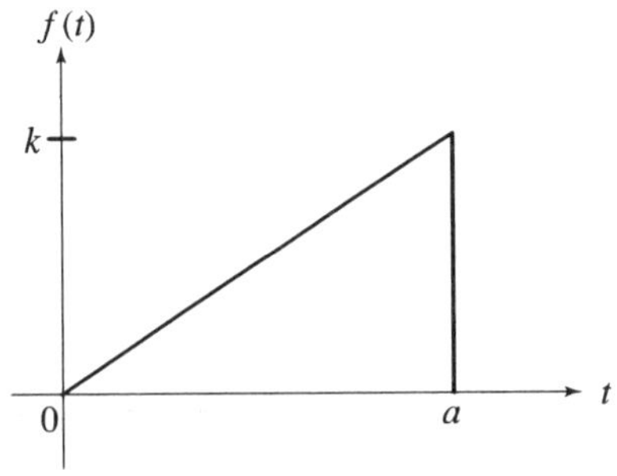

Figure 7-7 Sawtooth wave

Solution

Because $f(t)$ can be resolved into $f_1(t), f_2(t), f_3(t)$ as in Figure (7-8),

$$f(t) = f_1(t) + f_2(t) + f_3(t)$$

Where,

$$f_1(t) = \frac{k}{a} t u_o(t)$$

$$f_2(t) = -k u_a(t)$$

$$f_3(t) = -\frac{k}{a}(t-a) u_a(t)$$

Therefore,

$$\begin{aligned} F(s) &= \mathcal{L}[f(t)] = \mathcal{L}[f_1(t)] + \mathcal{L}[f_2(t) + \mathcal{L}[f_3(t)] \\ &= \frac{k}{a}\mathcal{L}[t] - k\mathcal{L}[u_a(t)] - \frac{k}{a}\mathcal{L}[(t-a)u_a] \\ &= \frac{k}{as^2}[1-(as+1)e^{-as}] \end{aligned}$$

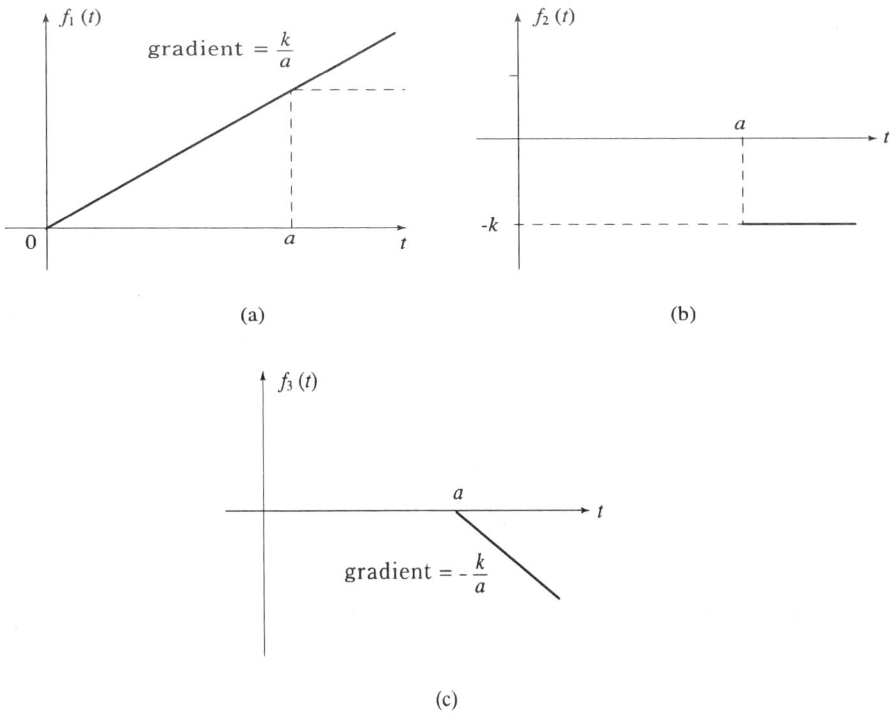

Figure 7-8 Resolved waveform of a sawtooth wave

Exercise 7-9

Evaluate Laplace transform of $F(s)=e^{-3s}/s^2$.

Solution

$$\mathcal{L}^{-1}\left[\frac{1}{s^2}\right]=t$$

Therefore, according to the theorem,

$$\mathcal{L}^{-1}[e^{-3s}/s^2]=(t-3)u_3(t)$$

Lastly, the Laplace transforms of periodic functions, frequently used in engineering, can be simply changed as described in the following theorem.

Theorem

If $f(t+nT)=f(t)$ $(n=1,2,\cdots\cdots)$, that is, $f(t)$ is a periodic function with the period T,

$$\mathcal{L}[f(t)]=\frac{1}{1-e^{-sT}}\int_0^T f(t)e^{-st}dt=\frac{F_1(s)}{1-2^{-sT}} \tag{7-48}$$

Because $f(t)$ is a periodic function with the period T,

$$\mathcal{L}[f(t)]=\int_0^\infty f(t)e^{-st}dt$$

$$=\sum_{n=0}^{\infty}\int_{nT}^{(n+1)T} f(t)e^{-st}dt]$$

Where,

$$t=\tau+nT$$

$$=\sum_{n=0}^{\infty}e^{-snT}\int_0^T f(\tau+nT)e^{-st}d\tau$$

$$=\frac{1}{1-e^{-sT}}\int_0^T f(\tau)e^{st}d\tau=\frac{F_1(s)}{1-e^{-sT}}.$$

Exercise 7-10

Evaluate Laplace transform of the sawtooth wave of the periodic function in Figure (7-9).

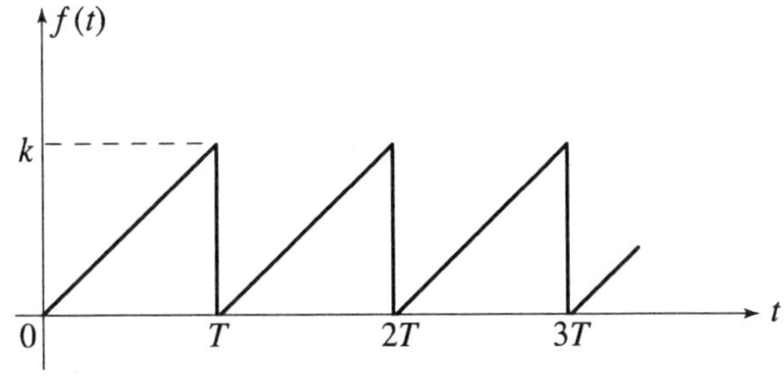

Figure 7-9 Sawtooth wave

Solution

If $0 < t < T$, $f(t) = \frac{k}{T}t$ and $f(t+T) = f(t)$.

According to partial integration,

$$\int_0^T te^{-st}dt = -\frac{t}{s}e^{-st}\Big|_0^T + \frac{1}{s}\int_0^T e^{-st}dt$$

$$= -\frac{T}{s}e^{-st} - \frac{1}{s^2}(e^{sT}-1).$$

Therefore, according to (7-48),

$$F(s) = \mathcal{L}[f(t)] = \frac{k}{Ts^2} - \frac{ke^{-sT}}{s(1-e^{-sT})}.$$

7. Convolution theorem

Let's find out the way to evaluate the response $h(t)$ for an input $f(t)$ with the response $g(t)$ for unit impulses. This theorem is useful in evaluating heat transfer, undulation and the flow of fluids. It can also be used in inverse Laplace transform. For an input $f(\tau)$ at $t=0$, the response or convolution $h(t)$ is defined as the following:

$$h(t)=\int_0^t f(\tau)g(t-\tau)d\tau \tag{7-49}$$

Where $f(\tau)=0$ when $\tau<0$, and $G(t-\tau)$ is a network function.

(7-49) is represented as follows:

$h(t)=f(t)*g(t)$

Theorem (Convolution theorem)

$$\mathcal{L}[h(t)]=\mathcal{L}[f(t)*(t)]=\mathcal{L}\left[\int_0^t f(\tau)g(t-\tau)d\tau\right]$$

$$=\mathcal{L}[f(t)]\mathcal{L}[g(t)]$$

$$H(s)=F(s)G(s) \tag{7-51}$$

This is called the convolution theorem.

Proof

$$\mathcal{L}[h(t)]=\mathcal{L}\left[\int_0^t f(\tau)g(t-\tau)d\tau\right]$$

$$=\mathcal{L}\left[\int_\tau^t f(\tau)g(t-\tau)u(t-\tau)d\tau\right]$$

$(\because u(t-\tau)=\begin{cases}1 & t\leqq t\\ 0 & \tau>t\end{cases}$, and therefore, the maximum integral value changes as $0\rightarrow 0, t\rightarrow\infty$.)

$$\mathcal{L}[h(t)]=\int_0^\infty\left[\int_0^\infty g(t-\tau)u(t-\tau)e^{-st}dt\right]f(\tau)d\tau$$

$$=\int_0^\infty\left[\int_0^\infty g(t-\tau)e^{-st}dt\right]f(\tau)d\tau$$

If $t=x+\tau$,

$$= \int_0^\infty \left[\int_0^\infty g(x)e^{-s(x+\tau)}dx\right] f(\tau)d\tau$$

$$= \int_0^\infty \left[\int_0^\infty g(x)e^{-sx}dx\right] f(\tau)e^{-s\tau}d\tau$$

$$= \left[\int_0^\infty f(\tau)e^{-s\tau}d\tau\right]\left[\int_0^\infty g(x)e^{-sx}dx\right]$$

$$= \mathcal{L}[f(t)]\mathcal{L}[g(t)] = F(s)G(s).$$

Exercise 7-11

Evaluate the following with the convolution theorem:

$y'' + \omega^2 y = w^2 f(t),\ y'(0) = y(0) = 0$

Solution

According to Laplace transform,

$$\mathcal{L}[y(t)] = Y(s),\ \mathcal{L}[f(t)] = F(s).$$

Therefore,

$$Y(s) = F(s)\frac{w^2}{s^2 + w^2} = F(s)G(s).$$

And

$$\mathcal{L}^{-1}\left[\frac{\omega^2}{s^2 + w^2}\right] = w\sin wt = g(t).$$

According to the convolution theorem,

$$\mathcal{L}(y) = \mathcal{L}(f)\mathcal{L}(g) = \mathcal{L}(f * g).$$

Therefore,

$$y(t) = f * g = \int_0^a f(x)g(t-\tau)t\tau$$

$$= w\int_0^t f(\tau)\sin w(t-\tau)d\tau.$$

Exercise 7-12

Evaluate $\mathcal{L}^{-1}\left[\frac{a^2}{(s^2+a^2)^2}\right]$ with the convolution theorem.

Solution

We cannot apply the partial fraction described in the previous section, but we can evaluate this with the convolution theorem.

$$H(s)=\frac{a^2}{(s^2+a^2)^2}\quad F(s)=\frac{a}{(s^2+a^2)}\quad G(s)=\frac{a}{s^2+a^2}$$

Therefore,

$$\mathcal{L}^{-1}\left[\frac{a}{(s^2+a^2)}\right]=\sin at=f(t)=g(t)$$

$$\begin{aligned}\mathcal{L}^{-1}\left[\frac{a^2}{(s^2+a^2)^2}\right]&=f(t)*g(t)\\&=\int_0^t f(\tau)g(t-\tau)d\tau\\&=\int_0^t \sin a\tau \sin a(t-\tau)d\tau\\&=\frac{1}{2}a(\sin at-at\cos at).\end{aligned}$$

Exercise 7-13

As in Figure 7-10 (a) and (b), there is an input $f(t)$ and the unit impulse response $g(t)$. Evaluate the response $h(t)$ for $f(t)$, using (a) Laplace transform and (b) convolution theorem.

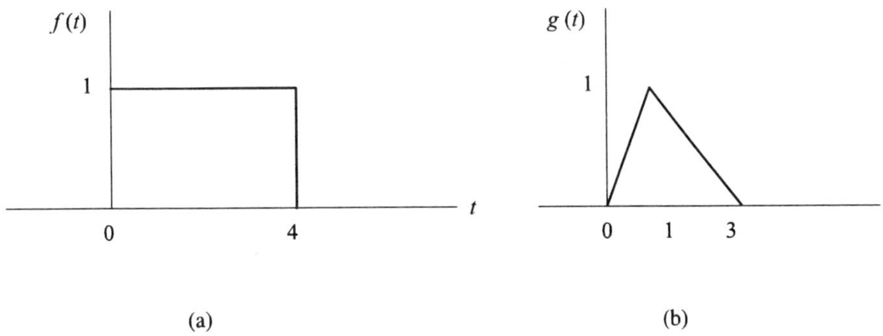

Figure 7-10 convolution of signal

Solution

(a) The figure is Laplace-transformed as follows:

$$F(s) = \frac{1}{s} - \frac{e^{-4s}}{s}$$

$$G(s) = \frac{1}{s^2} - \frac{3e^{-s}}{2s^2} + \frac{e^{-3s}}{2s^2}$$

According to (7-51),

$$H(s) = F(s)G(s) = \frac{1}{s^3} - \frac{3e^{-s}}{2s^3} + \frac{e^{-3s}}{2s^3} - \frac{e^{-4s}}{s^3} + \frac{3e^{-5s}}{2s^3} - \frac{e^{-7s}}{2s^3}.$$

If inverse-transformed, it becomes

$$h(t) = \mathcal{L}^{-1}[H(s)] = \frac{1}{2}t^2u_0(t) - \frac{3}{4}(t-1)^2u_1(t) + \frac{1}{4}(t-3)^2u_3(t)$$

$$-\frac{1}{2}(t-4)^2u_4(t) + \frac{3}{4}(t-5)^2u_5(t) - \frac{1}{4}(t-7)^2u_7(t).$$

(b) The function in the figure is represented with the following formula:

$f(t) = u_0(t) - u_4(t)$

$$g(t) = r(t) - \frac{3}{2}r(t-1) + \frac{1}{2}r(t-3)$$

Substitute this formula in (7-49) as follows:

$$h(t) = \int_0^t \left[r(\tau) - \frac{3}{2}r(\tau-1) + \frac{1}{2}r(\tau-3)\right] \cdot \left[u(t-\tau) - u(t-4-\tau)\right] d\tau$$

Where $u_a(t) = u(t-a)$.

$$= \int_0^t \left[r(\tau) - \frac{3}{2}r(\tau-1) + \frac{1}{2}r(\tau-3)\right] d\tau$$

$$- \int_0^{t-4} \left[r(\tau) - \frac{3}{2}r(\tau-1) + \frac{1}{2}r(\tau-3)\right] d\tau$$

$$= \int_0^t \tau d\tau - \frac{3}{2}\int_1^t (\tau-1)d\tau + \frac{1}{2}\int_3^t (\tau-3)d\tau$$

$$- \int_0^{t-4} \tau d\tau + \frac{3}{2}\int_1^{t-4} (\tau-1)d\tau - \frac{1}{2}\int_3^{t-4} (\tau-3)d\tau$$

$$= \frac{1}{2}\tau^2 u_0(t) - \frac{3}{4}(t-1)^2 u_1(t) + \frac{1}{4}(t-3)^2 u_3(t)$$

$$- \frac{1}{2}(t-4)^2 u_4(t) + \frac{3}{4}(t-5)^2 u_5(t) - \frac{1}{4}(t-7)^2 u_7(t)$$

The response may be drawn roughly as shown in Figure (7-11).

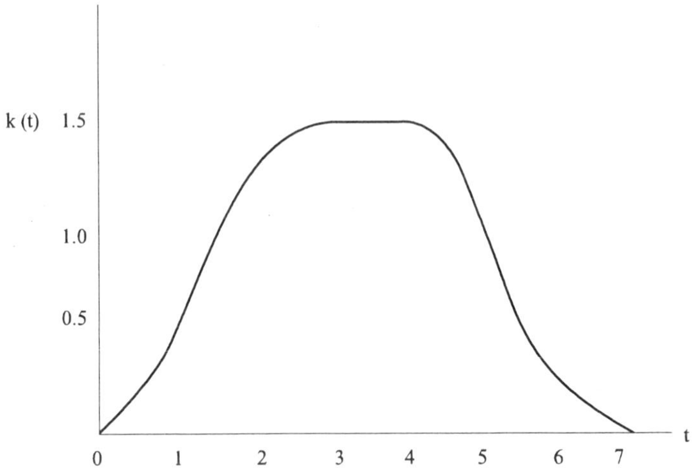

Figure 7-11 response function

Chapter 8

Fourier Series

1. Periodic function

If the period of a function $f(t)$ is T, the function $f(t)$ that satisfies the following condition is called a periodic function:

$$f(t) = f(t+T) \tag{8-1}$$

Figure (8-1) shows some functions that satisfy the above condition.

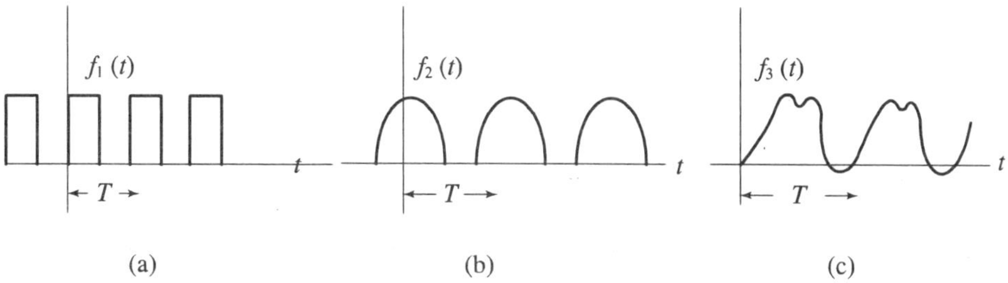

Figure 8-1 Periodic functions with the period of T

The following function, composed of two sine functions $f_1(t) = 3\sin\omega t$, $f_2(t) = \sin 2\omega t$, is a periodic function with the period of $2\pi/\omega = T$:

$$f(t) = 3\sin\omega t + \sin 2\omega t$$

Figure (8-2) shows $f_1(t)$, $f_2(t)$ and $f(t)$, where $f(t)$ is a non-sine wave with the period of $\frac{2\pi}{\omega}$.

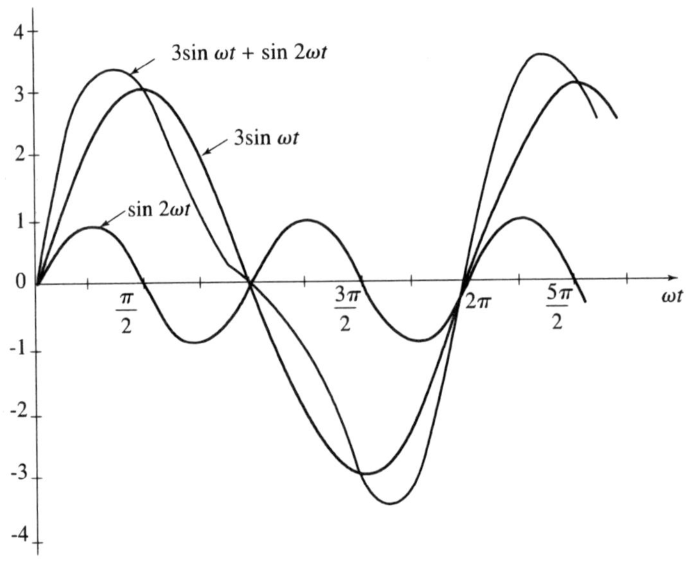

Figure 8-2 Sum of $3\sin\omega t$ and $\sin 2\omega t$

2. Fourier series

A periodic function may be expanded as a series of harmonic sine function. If $f(t)$ is a periodic functions, and satisfies the condition of Dirichlet, it can be expanded as follows:

$$f(t) = a_0 + a_1\cos\omega t + a_2\cos 2\omega t + a_3\cos 3\omega t + \cdots + a_n\cos n\omega t$$
$$+ b_1\sin\omega t + b_2\sin 2\omega t + b_3\sin\omega_3 t + \cdots + b_n\sin n\omega t$$

$$= a_0 + \sum_{n=1}^{\infty} a_n \cos n\omega t + \sum_{n=1}^{\infty} b_n \sin n\omega t \qquad (8\text{-}2)$$

Formula (8-2) is called a Fourier series. Dirichlet conditions are as follows:

(A) There must be a limited number of break points in a period.

(B) There must be a limited number of maximum points and minimum points in a period.

(C) $\int_0^t |f(t)|\, dt < \infty$ (Absolute convergence)

Fourier series applies only to those periodic function that satisfies above conditions. Most non-sinusoidal waves that concern us in engineering satisfy these conditions.

If a non-sinusoidal wave is given, the angular frequency ω can be found based on the period. To do so, one must evaluate each coefficient of Fourier series (this is called a Fourier coefficient).

The Fourier series formula (8-2) can be represented with sine terms only by combining cos terms and sin terms of the same frequency.

$$a_n \cos n\omega t + b_n \sin n\omega t = \sqrt{a_n^2 b_n^2}\left(\frac{a_n}{\sqrt{a_n^2 + b_n^2}} \cos n\omega t + \frac{b_n}{\sqrt{a_n^2 + b_n^2}} \sin n\omega t\right)$$

$$c_n \sin(n\omega t + \theta) \qquad (8\text{-}3)$$

Where,

$$c_n = \sqrt{a_n^2 + b_n^2} \quad (n = 1,\ 2,\ 3, \cdots) \qquad (8\text{-}4)$$

$$\theta_n = \tan^{-1}\frac{a_n}{b_n} \quad (n = 1,\ 2,\ 3, \cdots) \qquad (8\text{-}5)$$

Therefore, (8-2) can be written as follows:

$$f(t) = a_0 + \sum_{n=1}^{\infty} c_n \sin(n\omega t + \theta_n) \tag{8-6}$$

As we can see in (8-6), a non-sinusoidal periodic function can be divided into a constant term (DC component), a sinusoidal wave (fundamental wave) that has the same frequency as the periodic function, and a sinusoidal wave (harmonics) with a frequency that is integer times the frequency of the periodic function. This process is called Fourier analysis. (8-4) and (8-5) show the amplitude and phase angle of each component wave.

The graphs showing magnitude of c_n (amplitude) and θ_n (phase angle) for $n\omega$ are referred to as an amplitude spectrum and a phase spectrum, respectively. They are called a frequency spectrum.

Figure (8-3) shows the graphs of Fourier series showing time function and frequency spectrum, respectively. Figure (8-3(a)) showing the waveform in time domain is useful in finding the change of waveform by time. The frequency spectrum (frequency domain analysis) in Figure (8-3(b)) and (c) is useful for understanding the change of amplitude and phase of each frequency.

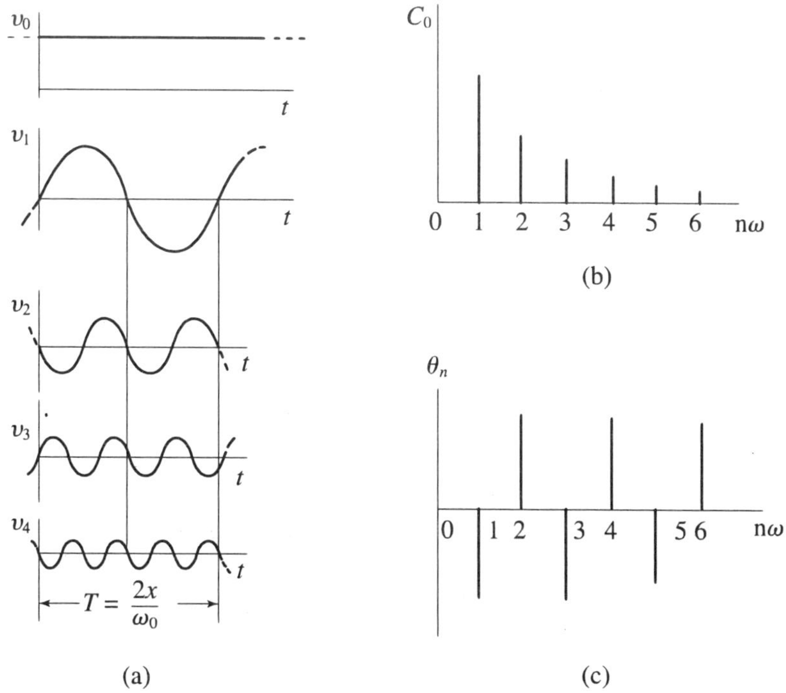

Figure 8-3 Fourier series graphs

3. Determination of Fourier coefficient

Determine from (8-2) the coefficient a_0 and the coefficient $a_n b_n$ having the positive values n and m.

1) Determination of a_0

To determine a_0, integrate both sides of (8-2) through the entire period from 0 to T.

$$\int_0^T f(t)dt = a_0 \int_0^T dt + \sum_{n=1}^{\infty} \int_0^T (a_n \cos n\omega t + b_n \sin n\omega t) dt$$

$$= a_0\int_0^T dt + \sum_{n=1}^{\infty} a_n \int_0^T \cos n\omega t + \sum_{n=1}^{\infty} b_n \int_0^T \sin n\omega t dt$$

$$= a_0 T + \sum_{n=1}^{\infty} a_n \frac{\sin n\omega t}{n}\Bigg|_0^T + \sum_{n=1}^{\infty} b_n \frac{-\cos n\omega t}{n}\Bigg|_0^T = a_0 T$$

$$\therefore a_0 = \frac{1}{T}\int_0^T f(t)dt \qquad (8\text{-}7)$$

We can find that a_0 is the average value for 1 period of the periodic function. Thus, this is also called a DC component.

2) Determination of a_n

To evaluate a_m, multiply (8-2) by $\cos m\omega t$, and integrate from 0 to T.

$$\int_0^T f(t)\cos m\omega t dt = a_o \int_0^T \cos m\omega t dt + \sum_{n=1}^{\infty} a_n \int_0^T \cos n\omega t \cdot \cos m\omega t dt$$

$$+ \sum_{n=1}^{\infty} b_n \int_0^T \sin n\omega t \cdot \cos m\omega t dt \qquad (8\text{-}8)$$

The 1st term of the right side is as follows:

$$\int_0^T \cos m\omega t dt = 0 \qquad (8\text{-}9)$$

The 2nd and 3rd terms are as follows:

$$\int_0^T \cos n\omega t \cdot \cos m\omega t dt = \frac{1}{2}\int_0^T [\cos(m+n)\omega t + \cos(m-n)\omega t]dt$$

$$= \begin{cases} 0; & (m \neq n) \\ \frac{T}{2}; & (m = n) \end{cases} \qquad (8\text{-}10)$$

$$\int_0^T \sin n\omega t \cdot \cos m\omega t dt = \frac{1}{2}\int_0^T [\sin(m+n)\omega t + \sin(m-n)\omega t]dt$$

$$= 0 \qquad (8\text{-}11)$$

Therefore,

$$\int_0^T f(t)\cos n\omega t dt = \frac{T}{2}a_n \qquad n=1,\ 2,\ 3,\cdots$$

$$\therefore a_n = \frac{2}{T}\int_0^T \cos n\omega t dt \qquad n=1,\ 2,\ 3,\cdots \tag{8-12}$$

3) Determination of b_n

Multiply both sides of (8-2) by $\sin m\omega t$, and integrate both sides from 0 to T.

$$\int_0^T f(t)\sin m\omega t dt = a_o\int_0^T \sin m\omega t dt + \sum_{n=1}^{\infty} a_n \int_0^T \cos n\omega t \cdot \sin m\omega t dt$$

$$+\sum_{n=1}^{\infty} b_n \int_0^T \sin n\omega t \cdot \sin m\omega t dt \tag{8-13}$$

The 1st term and 2nd term of the right side are as explained above, and the 3rd term is as follows:

$$\int_0^T \sin n\omega t \cdot \sin m\omega t dt = \frac{1}{2}\int_0^T [\cos(m-n)\omega t + \cos(m+n)\omega t]dt$$

$$= \begin{cases} 0, & (m \neq n) \\ \frac{T}{2}, & (m = n) \end{cases} \tag{8-14}$$

Therefore,

$$\int_0^T f(t)\sin n\omega t dt = b_n \frac{T}{2}$$

$$\therefore b_n = \frac{2}{T}\int_0^T f(t)\sin n\omega t dt \qquad n=1,\ 2,\ 3,\cdots \tag{8-15}$$

We can evaluate Fourier coefficients a_0, a_n, b_n as described above.

Exercise 8-1

Expand Fourier series for the pulse in Figure (8-4).

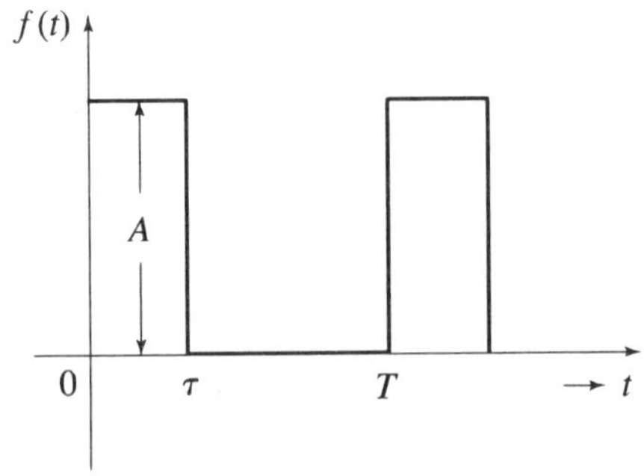

Figure 8-4

Solution

$$f(t)=\begin{cases} A & 0\leq t\leq \tau \\ 0 & \tau\leq t\leq T \end{cases}$$

From (8-7)

$$a_0=\frac{1}{T}\int_0^T f(t)=\frac{1}{T}\int_0^\tau Adt=\frac{1}{T}A\tau$$

From (8-11)

$$a_n=\frac{2}{T}\int_0^T f(t)\cos n\omega t dt=\frac{2}{T}\int_0^\tau A\cos n\omega t dt$$

$$=\frac{2A}{2\pi/\omega}\left[\frac{1}{n\omega}sin n\omega t\right]_0^\tau=\frac{A}{\pi n}sin n\omega\tau$$

From (8-13)

$$b_n=\frac{2}{T}\int_0^T f(t)\sin n\omega t dt=\frac{2}{T}\int_0^T A\sin n\omega t dt$$

$$=\frac{2A}{2\pi/\omega}\left[\frac{-1}{n\omega}cos n\omega t\right]_0^\tau=\frac{A}{\pi n}(1-\cos n\omega\tau)$$

Therefore,

$$f(t)=\frac{A\tau}{T}+\frac{A}{\pi}\Big\{\sin\omega\tau\cos\omega t+\frac{1}{2}\sin n\omega\tau\cos 2\omega t+\cdots+(1-\cos\omega\tau)$$

$$\sin\omega t+\frac{1}{2}(1-\cos 2\omega\tau)\sin 2\omega t+\cdots\Big\}$$

Exercise 8-2

Expand Fourier series for the standing wave in Figure (8-5).

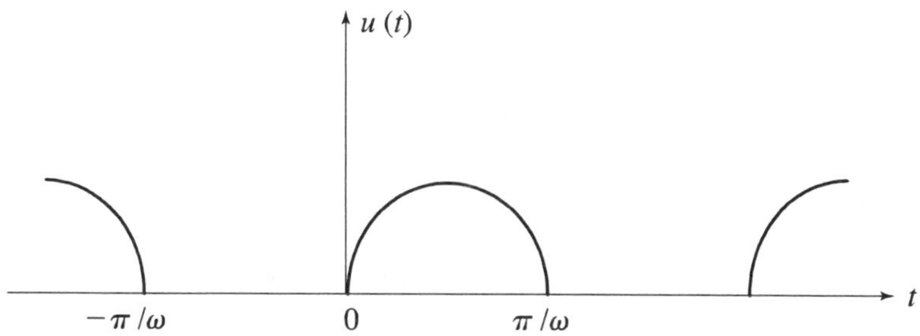

Figure 8-5

Solution

$$u(t)=\begin{cases}0 & -T/2<t<0\\ E\sin\omega t & 0<t\leq T/2\end{cases}$$

The frequency is $T=2\pi/\omega$.

When $-T/2<t<0$, $u=0$. Therefore,

$$a_0=\frac{\omega}{2\pi}\int_0^{\pi/\omega}E\sin\omega t\,dt=\frac{E}{\pi}$$

$$a_n=\frac{\omega}{\pi}\int_0^{\pi/\omega}E\sin\omega t\cos n\omega t\,dt$$

$$=\frac{\omega E}{2\pi}\int_0^{\pi/\omega}[\sin(1+n)\omega t+\sin(1-n)\omega t]dt$$

$$= \frac{\omega E}{2\pi}\left[-\frac{\cos(1+n)\omega t}{(1+n)\omega} - \frac{-\cos(1-n)\omega t}{(1-n)\omega}\right]_0^{\pi/\omega}$$

$$= \frac{E}{2\pi}\left[-\frac{\cos(1+n)\pi+1}{1+n} + \frac{-\cos(1-n)\pi+1}{1-n}\right]_0^{\pi/\omega}, \quad n \neq 1$$

If n is an odd number, $a_n = 0$. If n is an even number,

$$a_n = \frac{E}{2\pi}\left(\frac{2}{1+n} + \frac{2}{1-n}\right) = -\frac{2E}{(n+1)(n-1)\pi}.$$

In the same manner,

$$b_1 = \frac{E}{2},$$

$$b_n = 0 \quad (n=2,\ 3, \cdots).$$

Therefore,

$$u(t) = \frac{E}{\pi} + \frac{E}{2}\sin\omega t - \frac{2E}{\pi}\left(\frac{1}{1\cdot 3}\cos 2\omega t + \frac{1}{3\cdot 5}\cos 4\omega t + \cdots\right).$$

---

4. Symmetry

If a given function is symmetric, before determining Fourier coefficients, it is possible to expand Fourier series with either a_n or b_n. In this way, the calculation can be very simple.

1) Half-wave symmetry

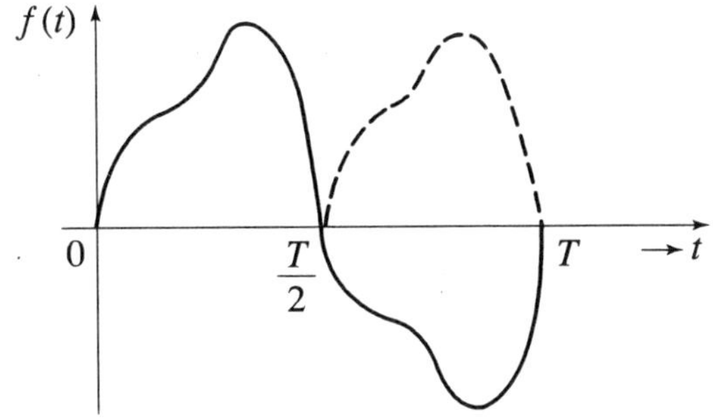

Figure 8-6 Half-wave symmetry

As in the case of Figure (8-6), if a wave switches between positive and negative to the same height for every half-period, and the positive half-wave corresponds to the negative half-wave when it is transferred in parallel by 180° to the horizontal axis and rotated by 180°, the wave is called a half-wave symmetry or symmetrical wave.

A half-wave symmetry is represented as follows:

$$f(t) = -f(t+\frac{T}{2}) \tag{8-16}$$

Because the area is the same between the positive half-wave and the negative half-wave, $a_0 = 0$.

Therefore, $f(t)$ is as follows:

$$f(t) = \sum_{n=1}^{\infty} a_n \cos n\omega t + \sum_{n=1}^{\infty} b_n \sin n\omega t \tag{8-17}$$

And $f(t+\frac{T}{2})$ is represented as follows:

$$f(t+\frac{T}{2}) = \sum_{n=1}^{\infty} a_n \cos n(\omega t+\frac{T}{2}) + \sum_{n=1}^{\infty} b_n \sin n(\omega t+\frac{T}{2})$$

where,

$$\sin n(\omega t+\frac{T}{2})=\sin(n\omega t+n\frac{T}{2})=(-1)^n\sin n\omega t$$

$$\cos n(\omega t+\frac{T}{2})=\cos(n\omega t+n\frac{T}{2})=(-1)^n\cos n\omega t$$

In order for (8-16) to be valid, n must be an odd number. Therefore, a symmetrical wave has the frequency component of multiples of odd numbers only.

Because $f(t)=-f(t+\frac{T}{2})$, in order to evaluate a_n, b_n, integrate for the half period only, and then, double the result.

$$a_n=\frac{4}{T}\int_0^{\frac{T}{2}}f(t)\cos n\omega t\,dt \quad (n \text{ is an odd number}) \tag{8-18}$$

$$b_n=\frac{4}{T}\int_0^{\frac{T}{2}}f(t)\sin n\omega t\,dt \quad (n \text{ is an odd number}) \tag{8-19}$$

2) Sine symmetry (odd function wave)

As in the case of Figure (8-7), if a positive half-wave corresponds to the negative half-wave when it is rotated by 180° to the vertical axis, and then to the horizontal axis, it is called a sine symmetry.

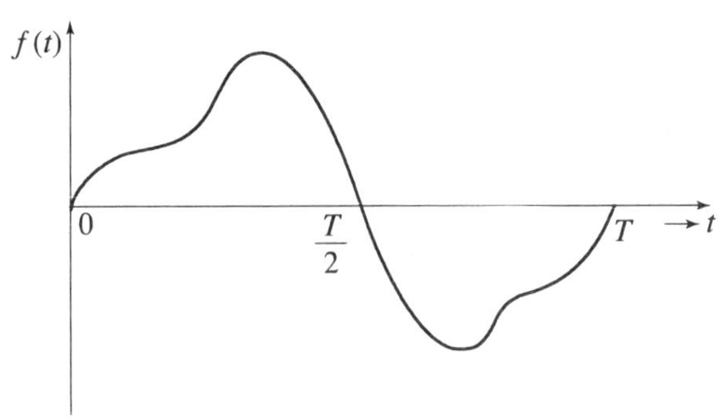

Figure 8-7 Sine symmetry

A sine symmetry is represented as the following:

$$\left.\begin{array}{r} f(t) = -f(t) \\ \text{or} \\ f(t) = -f(T-t) \end{array}\right\} \tag{8-20}$$

$f(t)$ is an odd function of t. Because the average value for a period is 0, $a_0 = 0$. When substituting $-\omega t$ in (8-2),

$$\cos n(-\omega t) = \cos n\omega t,\ \ \sin n(-\omega t) = -\sin\omega t$$

Therefore,

$$-f(-t) = -\left\{\sum_{n=1}^{\infty} a_n \cos n(-\omega t) + \sum_{n=1}^{\infty} b_n \sin n(-\omega t)\right\}$$

In order for the conditional expression (8-2) to be valid, only sine terms must exist.

Therefore, $a_0 = 0$ and $a_n = 0$. The coefficient b_n evaluated by integrating from 0 to $\frac{T}{2}$:

$$b_n = \frac{4}{T}\int_0^{\frac{T}{2}} f(t)\sin n\omega t\,dt \ \ (n \text{ is an odd number}) \tag{8-21}$$

3) Cosine symmetry (even function wave)

As in Figure (8-8), if a positive half-wave corresponds to the negative half-wave when it is rotated by 180° to the vertical axis, it is called a cosine symmetry.

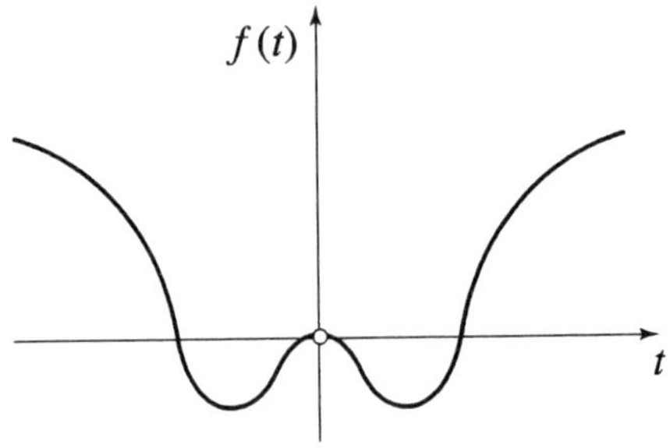

Figure 8-8 Cosine symmetry

$$\left.\begin{array}{c} f(t)=-f(t) \\ \text{or} \\ f(t)=-f(T-t) \end{array}\right\} \tag{8-22}$$

$f(t)$ is an even function of t. When substituting $-\omega t$ in (8-2),

$$\cos n(-\omega t)=\cos n\omega t, \quad \sin n(-\omega t)=-\sin n\omega t$$

Therefore, (8-22) is as follows:

$$f(-t)=a_0+\sum_{n=1}^{\infty}a_n\cos n(-\omega t)+\sum_{n=1}^{\infty}b_n\sin n(-\omega t)$$

$$=a_0+\sum_{n=1}^{\infty}a_n\cos n\omega t-\sum_{n=1}^{\infty}b_n\sin n\omega t$$

In order for the conditional expression (8-22) to be valid, there must be no sine term.

Therefore, $b_n=0$. To evaluate the coefficients a_0 and a_n, integrate for the half period only, and then, double the result.

$$\left.\begin{array}{l} a_0=\dfrac{1}{T}\displaystyle\int_0^T f(t)dt=\dfrac{T}{2}\int_0^{\frac{T}{2}} f(t)dt \\ a_n=\dfrac{2}{T}\displaystyle\int_0^T f(t)\cos n\omega t dt=\dfrac{4}{T}\int_0^{\frac{T}{2}} f(t)\cos n\omega t dt \end{array}\right\} \tag{8-23}$$

Table (8-1) summarizes the above results.

Table 8-1 Summary of symmetry of periodic waveform

Type of symmetry	Half-symmetry	Odd function	Even function
Condition	$f(t)=-f(t\pm\frac{T}{2})$	$f(t)=-f(t)$	$f(t)=f(-t)$
Waveform	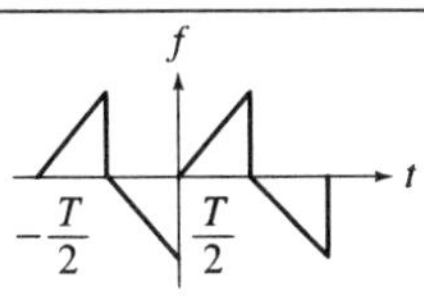	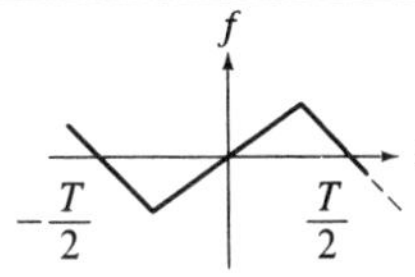	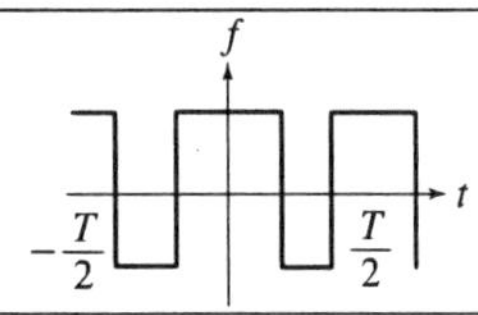
Characteristics	Odd harmonics only.	Sine term only.	Constant term and cosine term only.
$\frac{a_0}{2}$	0	0	$\frac{a_0}{2}=\frac{2}{T}\int_0^{\frac{T}{2}} f(t)dt$
a_n	$\frac{4}{T}\int_0^{\frac{T}{2}} f(t)\cos n\omega t dt$ $n=1,\ 3,\ 5,\cdots$	0	$\frac{4}{T}\int_0^{\frac{T}{2}} f(t)\sin n\omega t dt$ $n=1,\ 2,\ 3,\cdots$
b_n	$\frac{4}{T}\int_0^{\frac{T}{2}} f(t)\sin n\omega t dt$ $n=1,\ 3,\ 5,\cdots$	$\frac{4}{T}\int_0^{\frac{T}{2}} f(t)\cos n\omega t dt$ $n=1,\ 2,\ 3,\cdots$	0

Exercise 8-3

Evaluate Fourier series of the sawtooth wave as shown in Figure (8-9).

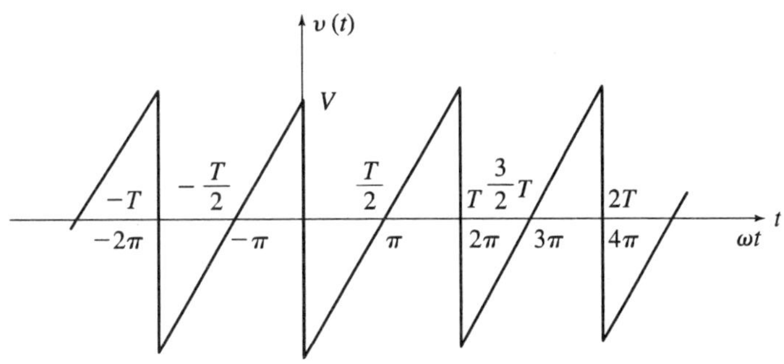

Figure 8-9 Sawtooth waveform

Solution

According to (8-15), b_n is calculated as the following:

$$b_n = \frac{2}{T}\int_0^T v(t)\sin n\omega t dt$$

The waveform is as follows:

$$v(t) = \frac{2V}{T}\left(t - \frac{T}{2}\right)$$

Therefore,

$$b_n = \frac{2}{T}\int_0^T \frac{2V}{T}\left(t - \frac{T}{2}\right)\sin n\omega t dt$$

$$= \frac{2}{T}\left[\int_0^T \frac{2V}{T} t \sin n\omega t dt - V\int_0^T \sin n\omega t dt\right]$$

$$= -\frac{2V}{n\pi}$$

Therefore,

$$b_1 = -\frac{2V}{\pi},\ b_2 = -\frac{2V}{\pi}\frac{1}{2},\ b_3 = -\frac{2V}{\pi}\frac{1}{3}, \cdots\cdots$$

Therefore, Fourier series of the waveform in Figure (8-9) is as follows:

$$v(t) = -\frac{2V}{\pi}(\sin\omega t + \frac{1}{2}sin2\omega t + \frac{1}{3}sin3\omega t + \cdots + \frac{1}{n}sinn\omega t + \cdots)$$

Exercise 8-4

Evaluate Fourier series of the standing wave as shown in Figure (8-10).

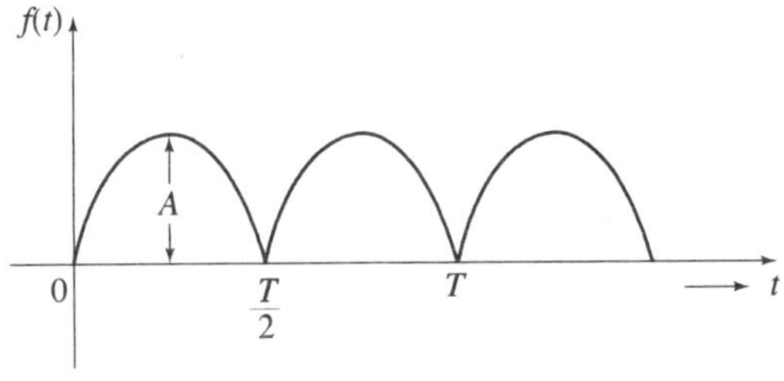

Figure 8-10 Full-wave rectification wave

$$f(t)=\begin{cases} A\sin\omega t & 0\leq t\leq \frac{T}{2} \\ -A\sin\omega t & \frac{T}{2}\leq t\leq T \end{cases}$$

Because $f(t)=-f(-t)$, it is cosine symmetric.

$$b_n=0$$

$$a_o=\frac{2}{T}\int_0^{\frac{T}{2}} f(t)dt=\frac{2}{T}\int_0^{\frac{T}{2}} A\sin\omega t dt=\frac{2A}{\pi}$$

$$a_n=\frac{4}{T}\int_0^{\frac{T}{2}} f(t)\cos n\omega t dt=\frac{4}{T}\int_0^{\frac{T}{2}} A\sin\omega t\cos n\omega t dt$$

$$=\begin{cases} 0 & n: odd\nu mber \\ -\frac{4A}{\pi}\frac{1}{(n-1)(n+1)} & n: even\nu mber \end{cases}$$

$$\therefore f(t)=\frac{4A}{\pi}\left(\frac{1}{2}-\frac{\cos 2\omega t}{1\cdot 3}-\frac{\cos 4\omega t}{3\cdot 5}-\frac{\cos 6\omega t}{5\cdot 7}\cdots\right)$$

Exercise 8-5

Evaluate Fourier series of the function illustrated in Figure (8-11).

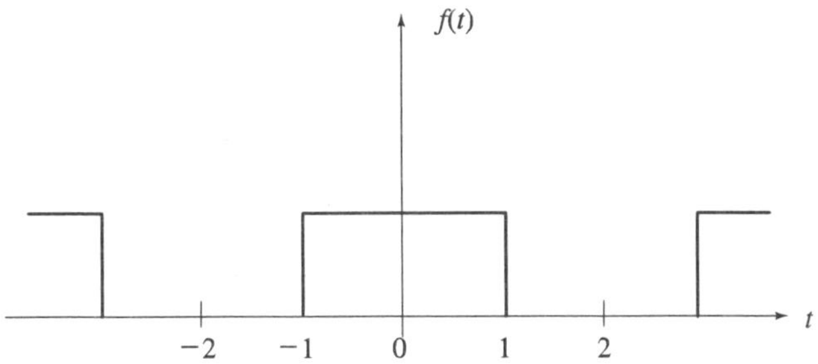

Figure 8-11 Square wave

Solution

In Figure (8-11),

$$f(t)=\begin{cases}0 & -2<t<-1\\ k & -1<t<1\\ 0 & 1<t<2\end{cases}$$

The period is $T=4$. From (8-7), (8-12) and (8-15),

$$a_0=\frac{1}{2}\int_0^2 f(t)dt=\frac{1}{2}\int_0^1 kdt=\frac{k}{2}$$

$$a_n=\int_0^2 f(t)\cos\frac{n\pi}{2}tdt=\int_0^1 k\cos\frac{n\pi}{2}tdt=\frac{2k}{n\pi}sin\frac{n\pi}{2}$$

Therefore,

$$a_n=\begin{cases}0 & : n=even\nu mber\\ \dfrac{2k}{n\pi} & : n=1,\ 5,9\\ \dfrac{-2k}{n\pi} & : n=3,\ 7,11\end{cases}$$

$$b_n=\int_0^2 f(t)\sin\frac{n\pi}{2}tdt=\int_0^1 k\sin\frac{n\pi}{2}tdt=0$$

Therefore,

$$f(t)=\frac{k}{2}+\frac{2k}{\pi}\left(\cos\frac{\pi}{2}t-\frac{1}{3}cos\frac{3\pi}{2}t+\frac{1}{5}cos\frac{5\pi}{2}t-\cdots\right)$$

Index

Engineering Maths for Applications

Published by O-Sung Media Ltd.
491-1 Chang 3-Dong, Dobong-Gu, Seoul, Korea
O-Sung and the device of a O-Sung are trademarks of O-Sung Media.

First Edition 2012. 08. 01.
Author Jin-Soo Han
Publisher Sung-Gul Oh
Tel. 02)997-0136 **Fax.** 02)997-0137
Registration 2000. 5. 12 (7-212)

ISBN 978-89-6628-046-9 Price : ₩**30,000**